Alexander Leonidovich Skubachevskii, Leonid Efimovich Rossovskii

Partial Differential Equations Theory

Also of Interest

Differential and Integral Calculus
Implicit Functions, Stieltjes Integrals and Curvilinear Integrals
Svetlin G. Georgiev, Khaled Zennir, 2025
ISBN 978-3-11-914462-9, e-ISBN (PDF) 978-3-11-221808-2

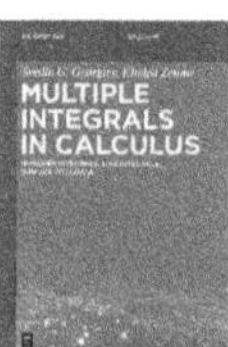

Multiple Integrals in Calculus
Improper Integrals, Line Integrals, Surface Integrals
Svetlin G. Georgiev, Khaled Zennir, 2025
ISBN 978-3-11-914345-5, e-ISBN (PDF) 978-3-11-221960-7

Differential Equations
A First Course on ODE and a Brief Introduction to PDE
Antonio Ambrosetti, Shair Ahmad, 2024
ISBN 978-3-11-118524-8, e-ISBN (PDF) 978-3-11-118567-5

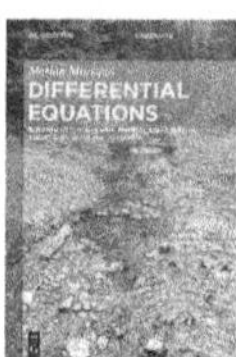

Differential Equations
Solving Ordinary and Partial Differential Equations with Mathematica®
Marian Mureşan, 2024
ISBN 978-3-11-141109-5, e-ISBN (PDF) 978-3-11-141139-2

Differential Geometry
Frenet Equations and Differentiable Maps
Muhittin E. Aydin, Svetlin G. Georgiev, 2024
ISBN 978-3-11-150089-8, e-ISBN (PDF) 978-3-11-150185-7

Differential Equations, Fourier Series, and Hilbert Spaces
Lecture Notes at the University of Siena
Raffaele Chiappinelli, 2023
ISBN 978-3-11-129485-8, e-ISBN (PDF) 978-3-11-130252-2

Alexander Leonidovich Skubachevskii,
Leonid Efimovich Rossovskii

Partial Differential Equations Theory

Sobolev Space, Weak Solution, Semigroup Theory, Fourier and Galerkin Methods

DE GRUYTER

Mathematics Subject Classification 2020
Primary: 35-01; Secondary: 46E35, 47D06

Authors
Alexander Leonidovich Skubachevskii
S.M. Nikol'skii Mathematical Institute
RUDN University
Miklukho-Maklaya str. 6
117198 Moscow
Russia
skubachevskiy_al@pfur.ru

Leonid Efimovich Rossovskii
S.M. Nikol'skii Mathematical Institute
RUDN University
Miklukho-Maklaya str. 6
117198 Moscow
Russia
rossovskiy_le@pfur.ru

ISBN 978-3-11-222962-0
e-ISBN (PDF) 978-3-11-222963-7
e-ISBN (EPUB) 978-3-11-222964-4

Library of Congress Control Number: 2025950406

Bibliographic information published by the Deutsche Nationalbibliothek
The Deutsche Nationalbibliothek lists this publication in the Deutsche Nationalbibliografie; detailed bibliographic data are available on the Internet at http://dnb.dnb.de.

Cover image: J614 / DigitalVision Vectors / Getty Images
Typesetting: VTeX UAB, Lithuania

www.degruyterbrill.com
Questions about General Product Safety Regulation:
productsafety@degruyterbrill.com

Introduction

This textbook is an introduction course to partial differential equations. Being concise, the book is intended for a wide audience but is primarily addressed to third-year students studying in the areas of “Mathematics” and “Applied Mathematics and Computer Science.” It is based on lectures given by A. L. Skubachevskii within the walls of the Moscow Aviation Institute and the Peoples’ Friendship University of Russia in the course of 30 years.

It is assumed that the student is familiar with the basics of calculus, linear algebra, ordinary differential equations, as well as function theory and functional analysis through classical university courses. However, in the first chapter, which is devoted to an introduction to the subject, the necessary brief information on the function theory and functional analysis is also given for the convenience of the reader since these disciplines are usually taught simultaneously with the course on partial differential equations.

The foundation for the material in the book is the functional approach associated with the concept of a weak solution and Sobolev spaces. This approach, first applied in [16] by S. L. Sobolev, serves as the basis for the modern theory of partial differential equations. S. L. Sobolev is also responsible for the emergence of new directions in analysis, such as the theory of distributions, the theory of function spaces, embedding theorems, etc. In this textbook, such an approach is demonstrated on second-order model equations: the Poisson equation, the heat equation, and the wave equation. However, it allows one to generalize the main results presented in the textbook to the case of more general equations of the corresponding type with variable coefficients.

The second chapter of this textbook is an introduction to the theory of Sobolev spaces. The third chapter is devoted to boundary-value problems for elliptic equations, and the fourth chapter deals with mixed problems and the Cauchy problem for equations of hyperbolic and parabolic types. The fifth chapter concerns the theory of semigroups of linear operators. At the end of the textbook, there is a series of exercises of various degrees of difficulty covering all the topics presented.

The theory of partial differential equations constructed on the basis of Sobolev spaces is presented in remarkable textbooks by O. A. Ladyzhenskaya [8], V. P. Mikhailov [11], and S. G. Mikhlin [12]. Book [1] served for many years as one of the main collections of exercises on partial differential equations. Undoubtedly, the more recent problem book [21] also deserves attention. We recommend using them for an in-depth study of the subject.

In addition to the sources mentioned above, there is a significant number of other notable books on the subject, see, for example, [2, 3, 7, 10, 13, 17, 18, 20, 22, 24, 25]. The textbook presented to the attention of readers uses most of the sources from the list of references. Its main difference is the presentation of the theory of semigroups and its application to evolutionary equations (thanks to this, mixed problems and the Cauchy problem are studied using a unified method), as well as the use of modern techniques, which made it possible to simplify a number of proofs. Note that this textbook is a new edition

https://doi.org/10.1515/9783112229637-202

of textbook [26, 27] published in English in a limited edition at the Peoples' Friendship University of Russia for the education of foreign students.

The authors are deeply grateful to Professor T. A. Suslina for her careful reading of the manuscript, useful advice, and correction of a number of inaccuracies. A number of valuable comments from Professors O. V. Besov, S. B. Kuksin, and A. I. Nazarov also contributed to the improvement of the book. The authors are grateful to Professors V. V. Kozlov and G. V. Demidenko for their attention to this work.

Contents

Notation

$\mathbb{R}$	set of all real numbers;
$\mathbb{R}_+$	positive real numbers;
$\mathbb{C}$	set of all complex numbers;
$\mathbb{R}^n$	n-dimensional real space;
$\mathbb{C}^n$	n-dimensional complex space;
$B_r(x^0)$	ball of radius r centered at x^0;
B_r	ball of radius r centered at 0;
$S_r(x^0)$	sphere of radius r centered at x^0;
σ_n	surface area of the unit sphere in $\mathbb{R}^n$;
$\square$	end of proof.

By $c, k, c_1, c_2, \dots, k_1, k_2, \dots$ we denote positive coefficients independent of the functions in inequalities.

https://doi.org/10.1515/9783112229637-204

1 Prerequisites

1 Subject of study

Main definitions and notation

Let us recall some of the well-known concepts of mathematical analysis.

Definition 1.1. A *partial differential equation (PDE)* is an equation that involves partial derivatives of the unknown function $u = u(x_1, \dots, x_n)$ of n independent real variables. The highest order of involved derivatives is called the *order* of the equation.

Definition 1.2. An ordered set $\alpha = (\alpha_1, \dots, \alpha_n)$ of nonnegative integers is called a *multiindex*. The notation $|\alpha| = \alpha_1 + \dots + \alpha_n$ is used throughout this book. For a given multiindex α, one can construct the linear operator of α-partial derivation

$$D^\alpha = \frac{\partial^{|\alpha|}}{\partial x_1^{\alpha_1} \dots \partial x_n^{\alpha_n}},$$

which differentiates a given function α_i times with respect to variable x_i, $i = 1, \dots, n$.

Remark 1.1. It is well known that if a function $u(x)$ is $|\alpha|$-continuously-differentiable in a neighborhood of a point x, then its mixed derivative $D^\alpha u(x)$ is independent of the differentiation order.

Definition 1.3. An open connected subset Q of the Euclidean space $\mathbb{R}^n$ is called a *domain*.

Definition 1.4. A point $x^0 \in \mathbb{R}^n$ is a *boundary point* of a domain Q if for all $\varepsilon > 0$ the ball $B_\varepsilon(x^0)$ intersects both Q and its complement $\mathbb{R}^n \setminus Q$. The set of all boundary points of Q is called the *boundary* of Q and is denoted by ∂Q. The set $\overline{Q} = Q \cup \partial Q$ is the *closure* of Q.

Definition 1.5. A partial differential equation is called *linear* if the unknown function and its partial derivatives are present in it in the form of a linear combination.

Each linear partial differential equation in a domain $Q \subset \mathbb{R}^n$ can be represented as follows:

$$\sum_{|\alpha| \le m} a_\alpha(x) D^\alpha u(x) = f(x) \quad (x \in Q).$$

In this representation, *coefficients* $a_\alpha(x)$ and the *right-hand side* $f(x)$ are given functions. We say that a partial differential equation is of order m at a point $x \in Q$ if

$$\sum_{|\alpha| = m} |a_\alpha(x)| \neq 0.$$

By using an alternative notation of partial derivatives in the case $m = 2$, one can rewrite a second-order linear equation as

https://doi.org/10.1515/9783112229637-001

$$\sum_{i,j=1}^{n} a_{ij}(x)u_{x_i x_j}(x) + \sum_{i=1}^{n} a_i(x)u_{x_i}(x) + a_0(x)u(x) = f(x) \quad (x \in Q), \\ \sum_{|\alpha|=m} |a_\alpha(x)| \neq 0 \quad (x \in Q). \tag{1.1}$$

Bearing Remark 1.1 in mind, we can assume that $a_{ij} = a_{ji}$ at each point of Q. From the perspective of applications, it is natural to consider only real-valued coefficients, even though it has to be noted that there are important examples of PDEs with complex-valued coefficients (e. g., the nonstationary Schrödinger equation). Equation (1.1) is called *homogeneous* if $f = 0$. It is called *nonhomogeneous* otherwise. The expression $\sum_{i,j=1}^{n} a_{ij}(x)u_{x_i x_j}(x)$ is the *leading term* of equation (1.1), while the other terms of the left-hand side are called *lower-order terms*.

Definition 1.6. Let Q be a domain and k be a nonnegative integer. By $C^k(Q)$, we denote the space of all k-times continuously differentiable functions in Q. By $C^k(\overline{Q})$, we denote the space consisting of all functions from $C^k(Q) \cap C(\overline{Q})$ such that all of their derivatives up to order k can be continuously extended to $\overline{Q}$, where $C(\overline{Q})$ is the set of all continuous in $\overline{Q}$ functions.

Definition 1.7. The set $\operatorname{supp}\varphi = \overline{\{x \in Q : \varphi(x) \neq 0\}}$ is called the *support* of a function $\varphi \in C(\overline{Q})$. If $\operatorname{supp}\varphi \subset Q$ is a compact set, then φ is called *compactly supported.* The set of all compactly supported functions from $C^k(\overline{Q})$ is denoted by $C_0^k(Q)$.

We also put

$$C^\infty(Q) = \bigcap_{k=0}^{\infty} C^k(Q), \quad C^\infty(\overline{Q}) = \bigcap_{k=0}^{\infty} C^k(\overline{Q}), \quad C_0^\infty(Q) = \bigcap_{k=0}^{\infty} C_0^k(Q).$$

Definition 1.8. We say that a boundary ∂Q *is of class* C^k if for each point $x^0 \in \partial Q$ there exist a number $i \in \{1,\dots,n\}$ and a ball $B_r(x^0)$ such that the set $\partial Q \cap B_r(x^0)$ can be uniquely projected along Ox_i-axis onto an $(n-1)$-dimensional domain $D = D(x^0, i, r)$ in the space of the variable $x' = (x_1,\dots,x_{i-1},x_{i+1},\dots,x_n)$ and

$$\partial Q \cap B_r(x^0) = \{x \in \mathbb{R}^n : x_i = \varphi(x'), x' \in D\},$$

where $\varphi \in C^k(\overline{D})$.

Example 1.1. Let $Q = \{x \in \mathbb{R}^2 : |x| < 1\}$. In vicinity of the points $(0,1)$ and $(0,-1)$, the boundary $\partial Q = \{x \in \mathbb{R}^2 : |x| = 1\}$ can be represented as graphs of the infinitely differentiable functions $x_2 = \sqrt{1-x_1^2}$ and $x_2 = -\sqrt{1-x_1^2}$, respectively. On the other hand, it is impossible to represent it as a graph of a function $x_1 = x_1(x_2)$. If instead we consider neighborhoods of $(1,0)$ and $(-1,0)$, then the boundary ∂Q can be represented as graphs of the functions $x_1 = \sqrt{1-x_2^2}$ and $x_1 = -\sqrt{1-x_2^2}$, respectively, but not as graphs of functions $x_2 = x_2(x_1)$. In vicinity of any other point of the boundary, both representations are possible, hence $\partial Q \in C^\infty$.

Definition 1.9. For domains $U, V \subset \mathbb{R}^n$ and a natural number k consider a bijection $f : U \to V$ such that all coordinate functions of the mappings f and f^{-1} are of class $C^k(U)$ and $C^k(V)$, respectively. Such bijection f is called a *C^k-diffeomorphism.*

If $\partial Q \in C^k$, then the mapping

$$y_1 = x_1 - x_1^0, \quad \dots, \quad y_{n-1} = x_{n-1} - x_{n-1}^0, \quad y_n = x_n - \varphi(x')$$

(we set $i = n$ in terms of Definition 1.8) is a C^k-diffeomorphism of some neighborhood U of x^0 onto a neighborhood V of the origin, and the image of $\partial Q \cap U$ is a part of the hyperplane $y_n = 0$. In other words, a boundary of class C^k can be locally straightened by an application of a C^k-diffeomorphism. In order to make sure of this statement, one could note that the Jacobian of such a mapping is equal to 1 and then apply the inverse function theorem.

Classification of second-order linear equations

Consider a matrix $A(x)$ of the leading term of equation (1.1):

$$A(x) = \begin{pmatrix} a_{11}(x) & \cdots & a_{1n}(x) \\ \vdots & \ddots & \vdots \\ a_{n1}(x) & \cdots & a_{nn}(x) \end{pmatrix}.$$

Since A is symmetric, it has exactly n real eigenvalues (taking multiplicity into account) at each point of the domain. Let us denote the number of positive eigenvalues by $n_+ = n_+(x)$, the number of negative eigenvalues by $n_- = n_-(x)$, and the multiplicity of zero eigenvalue by $n_0 = n_0(x)$. Note that $n_+ + n_- + n_0 = n$.

Definition 1.10. If $n_+ = n$ or $n_- = n$, then equation (1.1) is called *elliptic* at x. If $n_+ = 1$, $n_- = n-1$ or $n_+ = n-1$, $n_- = 1$, then equation (1.1) is called *hyperbolic* at x. If $n_+ + n_- = n$ and $n_+ \geq 2$, $n_- \geq 2$, then equation (1.1) is called *ultrahyperbolic* at x. Finally, if $n_0 > 0$, then equation (1.1) is called *parabolic* at x.

Remark 1.2. In general, numbers n_+, n_- and n_0 depend on x, hence the type of an equation may be different at different points of the domain.

Example 1.2. One of the most renowned examples of elliptic equations is the *Poisson equation* $-\Delta u = f$, where $\Delta = \partial^2/\partial x_1^2 + \dots + \partial^2/\partial x_n^2$ is the *Laplace operator*. This equation describes small elastic deformations of a membrane, stationary temperature distribution, electrostatic potential in a system of charged particles and other phenomena.

Example 1.3. The *wave equation* $u_{tt} - \Delta u = f$, where the unknown function $u = u(x_1, \dots, x_n, t)$ depends on n spacial variables $x_1, \dots, x_n$ and on time t, is a hyperbolic equation in each space-time domain in the $(n+1)$-dimensional space $\mathbb{R}^{n+1}_{x,t}$. The wave equation describes wave propagation in an elastic medium.

Example 1.4. The *heat equation* is the equation $u_t - \Delta u = f$ with respect to $u = u(x_1, \dots, x_n, t)$. It is of the parabolic type in any space-time domain of the $(n+1)$-dimensional space $R^{n+1}_{x,t}$. The heat equation describes the process of heat propagation. The heat equation is usually considered for $t > 0$.

Example 1.5. Here, we give a trivial example of an ultrahyperbolic equation:

$$u_{x_1x_1} + u_{x_2x_2} - u_{x_3x_3} - u_{x_4x_4} = 0, \quad u = u(x_1, \dots, x_4).$$

In all of the given examples, we are dealing with the *canonical form* of an equation, i. e., the form in which there are no mixed derivatives. One can try to simplify the structure of equation (1.1) using a suitable change of coordinates. Take any local C^2-diffeomorphism $x = x(y)$ and set $v(y) = u(x(y))$. Differentiating the composite function $u(x) = v(y(x))$, we obtain

$$u_{x_i} = \sum_{p=1}^{n} v_{y_p} \frac{\partial y_p}{\partial x_i}, \qquad u_{x_ix_j} = \sum_{p,q=1}^{n} v_{y_py_q} \frac{\partial y_p}{\partial x_i}\frac{\partial y_q}{\partial x_j} + \cdots$$

up to terms containing the first derivatives of the function v. Using these relations in equation (1.1), we obtain an equivalent equation for $v(y)$:

$$\sum_{p,q=1}^{n} b_{pq}(y) v_{y_py_q} + \text{lower-order terms} = g(y),$$

where $g(y) = f(x(y))$ and

$$b_{pq}(y) = \sum_{p,q=1}^{n} a_{ij}(x) \frac{\partial y_p}{\partial x_i}\frac{\partial y_q}{\partial x_j}\bigg|_{x=x(y)} \qquad (p, q = 1, \dots, n),$$

or, applying matrix notation,

$$B = JAJ^T. \tag{1.2}$$

Here, B is a matrix similar to A, composed of the coefficients b_{pq}, and J is the Jacobian matrix of the mapping $y = y(x)$. Since the matrix J is nonsingular, the numbers n_+, n_-, and n_0 for the matrix B at the point $y(x)$ are the same as for the matrix A at the point x. In other words, the above classification is invariant under diffeomorphisms. In addition, formula (1.2) can be used to reduce equation (1.1) to its canonical form. For a fixed point $x = x^0$, it is sufficient to find a nonsingular constant matrix J such that the matrix $JA(x^0)J^T$ turns out to be diagonal. After the linear change $y = Jx$, we obtain $b_{pq}(y^0) = 0$, $p \neq q$, in the new variables. Like any real symmetric matrix, $A(x^0)$ possesses an orthonormal basis composed of its eigenvectors in $\mathbb{R}^n$. Thus we can always take a matrix of these eigenvectors arranged in rows as J. In this case, the equation is reduced to its canonical form through an orthogonal transformation.

Note that if the coefficients a_{ij} in the leading part of equation (1.1) are constant, then the described change of variables brings the equation to its canonical form in the entire domain at once.

Example 1.6. Let us reduce the equation $u_{x_1x_2} = 0$ to its canonical form in $\mathbb{R}^2$.

To construct J, we find an orthonormal basis from the eigenvectors of the matrix

$$A = \begin{pmatrix} 0 & 1/2 \\ 1/2 & 0 \end{pmatrix}.$$

By writing the characteristic equation,

$$\det(A - \lambda E) = \begin{vmatrix} -\lambda & 1/2 \\ 1/2 & -\lambda \end{vmatrix} = \lambda^2 - \frac{1}{4} = 0,$$

we calculate the eigenvalues of the matrix A, $\lambda = \pm 1/2$. For the coordinates e_1 and e_2 of the eigenvector, we have

$$e_1 = e_2 = \frac{1}{\sqrt{2}} \quad \text{for } \lambda = \frac{1}{2},$$
$$e_1 = -\frac{1}{\sqrt{2}}, \quad e_2 = \frac{1}{\sqrt{2}} \quad \text{for } \lambda = -\frac{1}{2}.$$

Let us compile the desired matrix

$$J = \begin{pmatrix} 1/\sqrt{2} & 1/\sqrt{2} \\ -1/\sqrt{2} & 1/\sqrt{2} \end{pmatrix}$$

and the corresponding change of variables

$$y_1 = \frac{x_1 + x_2}{\sqrt{2}}, \quad y_2 = \frac{-x_1 + x_2}{\sqrt{2}},$$

representing the rotation of the coordinate axes by $\pi/4$ clockwise and reducing the equation $u_{x_1x_2} = 0$ to the form of a one-dimensional wave equation $v_{y_1y_1} - v_{y_2y_2} = 0$.

Statement of some problems of mathematical physics

Partial differential equations are rarely considered separately, by themselves. In applications, they are usually accompanied by *initial* and (or) *boundary conditions* that depict the physical meaning of the problem.

- *Cauchy problem for the wave equation*
  In the case of the Cauchy problem, the wave equation

$$u_{tt} - \Delta u = f(x,t) \quad (x \in \mathbb{R}^n,\ t > 0) \tag{1.3}$$

  is considered in the whole space $\mathbb{R}^n$, and the values of $u(x,t)$ and $u_t(x,t)$ are prescribed for $t = 0$,

$$u|_{t=0} = \varphi(x) \quad (x \in \mathbb{R}^n), \tag{1.4}$$

$$u_t|_{t=0} = \psi(x) \quad (x \in \mathbb{R}^n). \tag{1.5}$$

  Relations (1.4), (1.5) are called the *initial conditions*. The problem of finding a solution to equation (1.3) that satisfies conditions (1.4) and (1.5) is called the *Cauchy problem for the wave equation.*
- *Cauchy problem for the heat equation*
  In the Cauchy problem for the heat equation,

$$u_t - \Delta u = f(x,t) \quad (x \in \mathbb{R}^n,\ t > 0), \tag{1.6}$$

  considered in the whole space $\mathbb{R}^n$, only the values of $u(x,t)$ are specified at $t = 0$,

$$u|_{t=0} = \varphi(x) \quad (x \in \mathbb{R}^n). \tag{1.7}$$

  Problem (1.6), (1.7) is called *the Cauchy problem for the heat equation.*
- *Boundary-value problems for the Poisson equation*
  The Poisson equation

$$-\Delta u = f(x) \quad (x \in Q) \tag{1.8}$$

  in a (bounded) domain $Q \subset \mathbb{R}^n$ is often considered paired with one of the following three *boundary conditions*:

$$u|_{\partial Q} = \gamma(x) \quad (x \in \partial Q), \tag{1.9}$$

$$\left.\frac{\partial u}{\partial \nu}\right|_{\partial Q} = \gamma(x) \quad (x \in \partial Q), \tag{1.10}$$

  or

$$\left.\left(\frac{\partial u}{\partial \nu} + \sigma(x)u\right)\right|_{\partial Q} = \gamma(x) \quad (x \in \partial Q). \tag{1.11}$$

  Here, $\nu = \nu(x)$ is the unit outer normal vector to the boundary ∂Q at the point $x \in \partial Q$, $\partial u/\partial \nu$ is the normal derivative on ∂Q and $\sigma(x)$ is a given nonnegative function on ∂Q. Condition (1.9) prescribes values of the sought function $u(x)$ on ∂Q.

Problem (1.8), (1.9) is called *the Dirichlet problem* or *the first boundary-value problem*. In (1.10), the values of the normal derivative $\partial u/\partial \nu$ are specified on ∂Q. Problem (1.8), (1.10) is called *the Neumann problem* or *the second boundary-value problem*. Condition (1.11) connects the boundary values of the desired function and of its normal derivative. Problem (1.8), (1.11) is called *the Robin problem* or *the third boundary-value problem*. In all cases, the function γ is a given function defined on ∂Q. The homogeneous equation (1.8), $\Delta u = 0$, is called *the Laplace equation*.

- *Mixed problems for the wave equation*
  In the case where the wave equation is considered in a (bounded) spatial domain $Q \subset \mathbb{R}^n$, it is natural to state the *mixed* problem. In this formulation, the wave equation

$$u_{tt} - \Delta u = f(x,t) \quad ((x,t) \in \Omega_T = Q \times (0,T)) \tag{1.12}$$

is supplemented with both initial conditions

$$u|_{t=0} = \varphi(x) \quad (x \in Q), \tag{1.13}$$

$$u_t|_{t=0} = \psi(x) \quad (x \in Q), \tag{1.14}$$

and boundary conditions, which can be of the first type

$$u|_{\Gamma_T} = \gamma(x,t) \quad ((x,t) \in \Gamma_T = \partial Q \times (0,T)), \tag{1.15}$$

second type

$$\left.\frac{\partial u}{\partial \nu}\right|_{\Gamma_T} = \gamma(x,t) \quad ((x,t) \in \Gamma_T), \tag{1.16}$$

or third type

$$\left.\left(\frac{\partial u}{\partial \nu} + \sigma(x,t)u\right)\right|_{\Gamma_T} = \gamma(x,t) \quad ((x,t) \in \Gamma_T). \tag{1.17}$$

Here, $\nu = \nu(x)$ is the unit outer normal vector to the lateral surface Γ_T of the cylinder Ω_T at the point $(x,t) \in \Gamma_T$, $\partial u/\partial \nu$ is the normal derivative on Γ_T and $\sigma(x,t)$ is a given nonnegative function on Γ_T. Problem (1.12), (1.13), (1.14), (1.15) is called *the first mixed problem* for the wave equation. Problem (1.12), (1.13), (1.14), (1.16) is called *the second mixed problem* for the wave equation. Problem (1.12), (1.13), (1.14), (1.17) is called *the third mixed problem* for the wave equation. Here, γ is a given function on Γ_T.

- *Mixed problems for the heat equation*
  Similarly, for the heat equation

$$u_t - \Delta u = f(x,t) \quad ((x,t) \in \Omega_T = Q \times (0,T)) \tag{1.18}$$

in a (bounded) spatial domain $Q \subset \mathbb{R}^n$ problem (1.18), (1.13), (1.15) is called *the first mixed problem* for the heat equation. Problem (1.18), (1.13), (1.16) is called *the second mixed problem* for the heat equation. Problem (1.18), (1.13), (1.17) is called *the third mixed problem* for the heat equation.

Of course, the initial function φ and the boundary function γ appearing in the above problems have different physical meanings depending on the type of an equation and the type of a boundary condition.

2 Functional analysis

Lebesgue integral

In this section, we briefly outline the main facts related to the Lebesgue integral. To readers interested in proofs, we recommend books [5, Chapter V] and [14, Chapter 11]. The design of the Lebesgue integral has a number of advantages over the Riemann integral: it is more flexible, has a much wider scope of applicability and, most importantly, is better adapted to passages to limits.

Let us start by defining the Lebesgue measure in $\mathbb{R}^n$.

Definition 2.1. A set $A \subset \mathbb{R}^n$ is called *elementary* if it can be represented as a union of a finite number of coordinate parallelepipeds, i. e., sets of the form

$$P = \{(x_1, \dots, x_n) \in \mathbb{R}^n : a_i \leqslant x_i \leqslant b_i \ (a_i, b_i \in \mathbb{R}; i = 1, \dots, n)\}.$$

The case $a_i = b_i$ is allowed; some (or all) inequalities can be strict. The empty set is considered as a parallelepiped. Parallelepipeds in the definition of an elementary set can always be considered disjoint.

The family $\mathfrak{E}$ of all elementary sets forms a *ring*, which means that $A \cup B \in \mathfrak{E}$ and $A \setminus B \in \mathfrak{E}$ if $A \in \mathfrak{E}$ and $B \in \mathfrak{E}$.

Definition 2.2. Let us define a measure (volume) of a parallelepiped by the formula

$$\mu(P) = \prod_{i=1}^{n} (b_i - a_i).$$

For an elementary set $A = P_1 \cup \cdots \cup P_N$, we put

$$\mu(A) = \mu(P_1) + \cdots + \mu(P_N)$$

if the parallelepipeds $P_1, \dots, P_N$ are pairwise disjoint.

The value of $\mu(A)$ does not depend on how the elementary set A is decomposed into disjoint parallelepipeds, thus the function μ on the ring $\mathfrak{E}$ is well-defined. The function μ is *additive* on $\mathfrak{E}$: $\mu(A \cup B) = \mu(A) + \mu(B)$ for any disjoint elementary sets A and B.

Definition 2.3. For an arbitrary set $E \subset \mathbb{R}^n$, we define its *outer measure* by the formula

$$\mu^*(E) = \inf \sum_{k=1}^{\infty} \mu(I_k),$$

where the infimum is taken over all countable covers E by open parallelepipeds, $E \subset \bigcup_{k=1}^{\infty} I_k$. It may happen that $\mu^*(E) = +\infty$.

If $A \in \mathfrak{E}$, then $\mu^*(A) = \mu(A)$. In other words, the outer measure μ^* is an extension of the function μ from the ring $\mathfrak{E}$ to the set of all subsets of $\mathbb{R}^n$. In addition, the outer measure is *semiadditive*, which means that the inequality

$$\mu^*\left(\bigcup_{k=1}^{\infty} E_k\right) \leqslant \sum_{k=1}^{\infty} \mu^*(E_k)$$

holds for all $E_k \subset \mathbb{R}^n$.

Definition 2.4. For arbitrary sets $A, B \subset \mathbb{R}^n$, we define the "distance" $d(A, B)$ between A and B as the outer measure $\mu^*(S(A, B))$ of their symmetric difference $S(A, B) = (A \setminus B) \cup (B \setminus A)$.

Note that the value $d(A, B)$ has the following usual metric properties: $d(A, B) \geqslant 0$, $d(A, A) = 0$, $d(A, B) = d(B, A)$ and $d(A, B) \leqslant d(A, C) + d(B, C)$. However, the equality $d(A, B) = 0$ does not imply that $A = B$.

Definition 2.5. A set $A \subset \mathbb{R}^n$ is called *finitely measurable* if there exists a sequence $\{A_k\}$ of elementary sets such that $d(A_k, A) \to 0$ as $k \to \infty$.

A set is called *measurable* if it can be represented as a union of a countable collection of finitely measurable sets.

If a set A is finitely measurable, then $\mu^*(A) < +\infty$. This is not true for an arbitrary measurable set A, where $\mu^*(A)$ can equal $+\infty$.

Theorem 2.1. *The family $\mathfrak{M}$ of all measurable subsets of $\mathbb{R}^n$ forms a σ-ring, and the function $\mu^* : \mathfrak{M} \to [0, +\infty]$ is countably additive on $\mathfrak{M}$. In other words, for $A_k \in \mathfrak{M}$ $(k = 1, 2 \dots)$ we have*

$$\bigcup_{k=1}^{\infty} A_k \in \mathfrak{M} \quad \text{and} \quad A_1 \setminus A_2 \in \mathfrak{M},$$

and if, in addition, the measurable sets A_k are pairwise disjoint, then

$$\mu^*\left(\bigcup_{k=1}^{\infty} A_k\right) = \sum_{k=1}^{\infty} \mu^*(A_k).$$

The constructed *countably additive function* μ^* on the σ-ring $\mathfrak{M}$ of all measurable sets is called the *Lebesgue measure*. It is customary to write $\mu(A)$ instead of $\mu^*(A)$ if $A \in \mathfrak{M}$. Another common notation we use in the book is $|A|$.

Remark 2.1. If $A \subset \mathbb{R}^n$ is open or closed, then $A \in \mathfrak{M}$. If $A \in \mathfrak{M}$ and $\varepsilon > 0$, then there exists a closed set F and an open set G such that $F \subset A \subset G$ and

$$\mu(G \setminus A) < \varepsilon, \quad \mu(A \setminus F) < \varepsilon.$$

Another important property of the Lebesgue measure μ is its completeness: from the relations $B \subset A$, $A \in \mathfrak{M}$ and $\mu(A) = 0$, it follows that $B \in \mathfrak{M}$ and $\mu(B) = 0$.

Integration theory deals only with measurable sets. Throughout this section, E denotes a measurable subset of $\mathbb{R}^n$.

Definition 2.6. A function $f : E \to \mathbb{R}$ is called *measurable* if $f^{-1}(V) \in \mathfrak{M}$ for any open set $V \subset \mathbb{R}$.

The definition of a measurable function does not change if we require the measurability only of sets of the form $\{x \in E : f(x) > a\}$, $a \in \mathbb{R}$.

Obviously, every continuous function is measurable. The functions $|f|$, $f + g$, fg, $\max(f, g)$ and $\min(f, g)$ are measurable if f and g are measurable. The limit of a convergent sequence of measurable functions is a measurable function.

Definition 2.7. A function $s : E \to \mathbb{R}$ is called *simple* if it takes a finite number of values.

Assuming that the image of a simple function s consists of points $c_1, \ldots, c_N$, we can write this function in the form

$$s(x) = \sum_{i=1}^{N} c_i \chi_{E_i}(x),$$

where $E_i = \{x \in E : s(x) = c_i\}$, $\bigcup_{i=1}^{N} E_i = E$ and χ_{E_i} is the characteristic function of the set E_i. A simple function is measurable if and only if all sets E_i are measurable.

First, the Lebesgue integral is constructed for nonnegative functions. For a nonnegative measurable simple function s, the value

$$I(s) = \sum_{i=1}^{N} c_i \mu(E_i)$$

is defined. Note that $I(s)$ is either a nonnegative number or $+\infty$.

Taking an arbitrary nonnegative function $f(x)$, $x \in E$, we introduce the sets

$$E_{k,i} = \left\{x \in E : \frac{i-1}{2^k} \leqslant f(x) < \frac{i}{2^k}\right\} \quad (k = 1, 2 \ldots;\ i = 1, \ldots, k2^k),$$

$$F_k = \{x \in E : f(x) \geqslant k\} \quad (k = 1, 2, \ldots)$$

and a nondecreasing sequence

$$s_k(x) = \sum_{i=1}^{k2^k} \frac{i-1}{2^k}\chi_{E_{k,i}} + k\chi_{F_k}$$

of nonnegative simple functions converging to $f(x)$ on E. If f is measurable, then the functions s_k are also measurable. If f is bounded, then the sequence $s_k(x)$ converges to $f(x)$ uniformly on E.

Thus, given a nonnegative measurable function f on a measurable set E, a nondecreasing sequence $\{I(s_k)\} \subset [0, +\infty]$ is constructed.

Definition 2.8. The quantity

$$\int_E f(x)\,dx = \sup_k I(s_k)$$

is called the *Lebesgue integral* of the function f over the set E.

Remark 2.2. We have $\sup_k I(s_k) = \lim_{k\to\infty} I(s_k)$. At the same time, the supremum in Definition 2.8 does not change if it is taken over all measurable simple functions s such that $0 \leqslant s \leqslant f$ [14, Theorem 11.28]. If the function f is simple, then $\int_E f\,dx = I(f)$.

For an arbitrary real-valued measurable function f, we use the expansion $f = f^+ - f^-$, where functions $f^+ = \max(f, 0)$ and $f^- = -\min(f, 0)$ are nonnegative and measurable.

Definition 2.9. Let us introduce the Lebesgue integral of a real-valued measurable function f, setting

$$\int_E f\,dx = \int_E f^+\,dx - \int_E f^-\,dx,$$

if at least one of the integrals $\int_E f^+\,dx$, $\int_E f^-\,dx$ is finite. If both of them are finite, then the function f is called *Lebesgue integrable* on the set E. The set of all functions integrable on a measurable set E is called *the Lebesgue space* and is denoted by $L_1(E)$.

Remark 2.3. The construction introduced above has the following key property: $f \in L_1(E)$ if and only if $|f| \in L_1(E)$, and

$$\left|\int_E f\,dx\right| \leqslant \int_E |f|\,dx = \int_E f^+\,dx + \int_E f^-\,dx.$$

Let us proceed to the integration of complex-valued functions f on E.

Definition 2.10. A complex-valued function $f(x) = u(x) + iv(x)$, $x \in E$ is called measurable if the real-valued functions $u(x)$ and $v(x)$ are measurable.

If f is measurable, then the function $|f| = (u^2 + v^2)^{1/2}$ is measurable as well.

Definition 2.11. We say that a complex-valued function f is integrable on E and write $f \in L_1(E)$ if $u \in L_1(E)$ and $v \in L_1(E)$. And it is assumed that

$$\int_E f\,dx = \int_E u\,dx + i\int_E v\,dx.$$

Just as for real-valued functions, $f \in L_1(E)$ is equivalent to $|f| \in L_1(E)$.

Theorem 2.2. (a) *If a measurable function f is such that $|f(x)| \leqslant g(x)$ on E and $g \in L_1(E)$, then $f \in L_1(E)$ and*

$$\left|\int_E f\,dx\right| \leqslant \int_E |f|\,dx \leqslant \int_E g\,dx.$$

In particular, if f is measurable and bounded on E, i. e., there exists a constant $M > 0$ such that $|f(x)| \leqslant M$ for $x \in E$, and if $\mu(E) < +\infty$, then $f \in L_1(E)$ and $|\int_E f\,dx| \leqslant M\mu(E)$.

(b) *If $f \in L_1(E)$ and $\varepsilon > 0$, then there exists a number $\delta > 0$ such that*

$$\left|\int_{A\cap E} f\,dx\right| < \varepsilon$$

for any set $A \in \mathfrak{M}$ such that $\mu(A) < \delta$.

(c) *For any fixed function $f \in L_1(E)$, the complex-valued set function*

$$\varphi(A) = \int_{A\cap E} f\,dx \quad (A \in \mathfrak{M})$$

is countably additive on the σ-ring $\mathfrak{M}$ of all measurable subsets of $\mathbb{R}^n$.

(d) *If $f_1, f_2 \in L_1(E)$ and $c_1, c_2 \in \mathbb{C}$, then $c_1f_1 + c_2f_2 \in L_1(E)$ and*

$$\int_E (c_1f_1 + c_2f_2)\,dx = c_1\int_E f_1\,dx + c_2\int_E f_2\,dx.$$

Thus, $L_1(E)$ is a linear space, and $\int_E f\,dx$ is a linear functional on $L_1(E)$.

Property (b) is called the *absolute continuity* of the Lebesgue integral.

An important role in analysis is played by the following statement, known as *Lebesgue's dominated convergence theorem.*

Theorem 2.3. *If a function $f(x)$, $x \in E$, is the limit of a convergent sequence of measurable functions $f_k(x)$ and there exists an integrable function $g(x)$ such that*

$$|f_k(x)| \leqslant g(x) \quad (x \in E;\ k = 1, 2, \ldots),$$

then the function f is integrable as well and

$$\lim_{k\to\infty}\int_E f_k\,dx = \int_E f\,dx.$$

Remark 2.4. From the definition of the Lebesgue integral (Definitions 2.8, 2.9 and 2.11) it immediately follows that $\int_E f\,dx = 0$ for any measurable function f if $\mu(E) = 0$. Supplied with the additivity property, this leads to the equality

$$\int_A f\,dx = \int_B f\,dx$$

for a measurable function f and for all measurable sets A and B such that $\mu(S(A,B)) = 0$, where the integrals exist or do not exist simultaneously. In other words, when integrating, sets of zero measure can be neglected.

Definition 2.12. When a property holds for all points of the set E, except for a subset of measure zero, we say that the property holds *almost everywhere* on E or *for almost all* $x \in E$.

It is often enough to assume that a property holds almost everywhere. For example, the statement of Lebesgue's dominated convergence theorem remains true if the sequence $f_k(x)$ converges almost everywhere on E and the estimate $|f_k(x)| \leqslant g(x)$ holds almost everywhere on E.

Moreover, one can allow the functions to be defined not at every point, but almost everywhere on the set, or if necessary, change their values on subsets of zero measure without any influence on the integral. Therefore, in the integration theory, it is appropriate to understand a function as the class of equivalence of functions that coincide almost everywhere.

Remark 2.5. The Lebesgue integral is constructed according to the same scheme, preserving all the properties mentioned above in a much more general situation, namely for a numerical function f defined on an abstract *measurable space* X. The latter means that some σ-ring $\mathfrak{M}$ of distinguished subsets of X is fixed and there exists a countably-additive set function μ on $\mathfrak{M}$ with values from $[0, +\infty]$, called a *measure*. This is a significant advantage of the Lebesgue integral. The case of the Lebesgue measure in $\mathbb{R}^n$ was chosen for clarity and in connection with the main content of the book, where a bounded domain Q in $\mathbb{R}^n$ or the space $\mathbb{R}^n$ itself appears as E.

This paragraph ends with *Fubini's theorem*—a key statement in the theory of the Lebesgue multiple integral.

Let us represent $\mathbb{R}^{n+m}$ as a Cartesian product of $\mathbb{R}^n_x \times \mathbb{R}^m_y$. Suppose that a set $E \subset \mathbb{R}^{n+m}$ is measurable in $\mathbb{R}^{n+m}$, i. e., with respect to the Lebesgue measure constructed in a similar way for $\mathbb{R}^{n+m}$. Then the section $E_x = \{y \in \mathbb{R}^m : (x,y) \in E\}$ is a measurable set in $\mathbb{R}^m$ for almost all $x \in \mathbb{R}^n$, and the section $E_y = \{x \in \mathbb{R}^n : (x,y) \in E\}$ is measurable in $\mathbb{R}^n$ for almost all $y \in \mathbb{R}^m$.

Let $f = f(x,y)$ be a measurable complex-valued function on E. Then it is a measurable function of y for almost all $x \in \mathbb{R}^n$ (such that the set E_x is nonempty), and also a measurable function of x for almost all $y \in \mathbb{R}^m$ (such that the set E_y is nonempty). The notation $f(x,y) = f_x(y)$ or $f(x,y) = f_y(x)$ is used depending on whether x or y is being fixed. The integrals $\int_{E_x} f_x(y)\,dy$ and $\int_{E_y} f_y(x)\,dx$ are considered equal to zero in the case where the corresponding integration set is empty.

Theorem 2.4. (a) *If $f \in L_1(E)$, then $f_x \in L_1(E_x)$ for almost all $x \in \mathbb{R}^n$, $f_y \in L_1(E_y)$ for almost all $y \in \mathbb{R}^m$, the functions*

$$\varphi(x) = \int_{E_x} f_x(y)\,dy, \quad \psi(y) = \int_{E_y} f_y(x)\,dx$$

belong to $L_1(\mathbb{R}^n)$ and $L_1(\mathbb{R}^m)$, respectively, and

$$\int_E f(x,y)\,dxdy = \int_{\mathbb{R}^n} \varphi(x)\,dx = \int_{\mathbb{R}^m} \psi(y)\,dy.$$

(b) *If at least one of the functions*

$$\varphi^*(x) = \int_{E_x} |f_x(y)|\,dy, \quad \psi^*(y) = \int_{E_y} |f_y(x)|\,dx$$

is integrable, then $f \in L_1(E)$.

Combining the statements in Fubini's theorem, we arrive at the following useful result: if

$$\int_{\mathbb{R}^n} dx \int_{E_x} |f(x,y)|\,dy < +\infty \quad \text{or} \quad \int_{\mathbb{R}^m} dy \int_{E_y} |f(x,y)|\,dx < +\infty,$$

then both iterated integrals of the function f are finite and equal to the double integral,

$$\int_E f(x,y)\,dxdy = \int_{\mathbb{R}^n} dx \int_{E_x} f(x,y)\,dy = \int_{\mathbb{R}^m} dy \int_{E_y} f(x,y)\,dx.$$

In other words, the order of integration in an iterated integral can be changed provided that at least one of the iterated integrals of $|f|$ is finite.

Normed and unitary spaces

In this section, for the convenience of the reader, initial information about metric spaces, normed spaces and inner product spaces is collected. This material can be found in any textbook on functional analysis, including [4, 5, 15].

Definition 2.13. A *metric space* is a set X paired with a *distance function* ρ (in other words, a *metric*) with the following properties:
(a) $0 \leqslant \rho(x,y) < \infty$ for all $x, y \in X$.
(b) $\rho(x,y) = 0$ if and only if $x = y$.
(c) $\rho(x,y) = \rho(y,x)$ for all $x, y \in X$.
(d) $\rho(x,y) \leqslant \rho(x,z) + \rho(z,y)$ for all x, y and $z \in X$.

Relation (d) is called the *triangle inequality*. For $x \in X$ and $r > 0$, *an open ball* $B_r(x)$ of radius r with center x is the set $\{y \in X : \rho(x,y) < r\}$. A subset of a metric space is called *open* if it can be represented as a union of a certain set of open balls. Every open set containing the point x is called a *neighborhood* of the point x.

Definition 2.14. We say that a sequence x_n *converges* to x in the metric space X and write $x_n \to x$ if $\rho(x_n, x) \to 0$ as $n \to \infty$. A sequence x_n is called a *Cauchy sequence* in X if $\rho(x_n, x_m) \to 0$ as $n, m \to \infty$.

Definition 2.15. A metric space is called *complete* if every Cauchy sequence converges to an element of this space.

Definition 2.16. A metric space X is called *separable* if there exists a countable set $E \subset X$ that is dense in X. The latter means that every ball in X contains points from E.

Definition 2.17. A set in a metric space is *closed* if its complement is open. The *closure* $\overline{E}$ of a set E is the intersection of all closed sets containing E. A point $x \in X$ is called a *limit point* of a set $E \subset X$ if any neighborhood of the point x contains at least one point of the set E different from x. Equivalently, the closure of a set E is the union of E and the set of all limit points of the set E. A subset of a metric space is *bounded* if it is contained in some ball. A subset K of a metric space is called *compact* if any covering of K by open sets contains a finite subcovering. This is equivalent to the fact that any sequence in K contains a subsequence converging to some point in the set K. A subset K of a metric space is called *precompact* if its closure is compact.

Every compact set is closed and bounded but the converse is, generally speaking, not true (for sets lying in $\mathbb{R}^n$ or $\mathbb{C}^n$, compactness is equivalent to closedness and boundedness).

Definition 2.18. A mapping f of a metric space X into a metric space Y is called *continuous* if the preimage $f^{-1}(U)$ of any open set U in Y is an open set in X. This is equivalent to saying that $x_n \to x$ in X implies $f(x_n) \to f(x)$ in Y.

Definition 2.19. A set V is called a *complex linear space* and its elements are called *vectors* if it is paired with two operations, *addition* and *multiplication by complex numbers*, satisfying the properties listed below.

To every pair of vectors x and y there corresponds a vector $x+y$, in such a way that $x+y=y+x$ and $x+(y+z)=(x+y)+z$; V contains a unique vector 0 (*null vector*) such that $x+0=x$ for all $x \in V$; and to each $x \in V$ there corresponds a unique vector $-x$ such that $x+(-x)=0$.

To each pair (α, x), where $x \in V$ and $\alpha \in \mathbb{C}$, there is associated a vector $\alpha x \in V$, in such a way that $1x = x$, $\alpha(\beta x) = (\alpha\beta)x$, and the distributive laws $\alpha(x+y) = \alpha x + \alpha y$ and $(\alpha+\beta)x = \alpha x + \beta x$ hold.

Replacing the field of complex numbers $\mathbb{C}$ with the field of real numbers $\mathbb{R}$, we obtain a definition of a *real linear space*.

Definition 2.20. A finite system of vectors $x_1, x_2, \dots, x_n$ is called *linearly independent* if the relation $\alpha_1 x_1 + \cdots + \alpha_n x_n = 0$ implies $\alpha_1 = \cdots = \alpha_n = 0$. An infinite system is *linearly independent* if all its finite subsystems are linearly independent.

Definition 2.21. A *linear operator* acting from a linear space V to a linear space V_1 is a mapping A from V to V_1 such that

$$A(\alpha x + \beta y) = \alpha Ax + \beta Ay$$

for all vectors x and $y \in V$ and for all scalars α and β.

In the case where $V_1 = \mathbb{C}$ (or $V_1 = \mathbb{R}$), A is called a *linear functional*.

Definition 2.22. A linear space V is called *normed* if each vector $x \in V$ is associated with a number $\|x\|$, called its *norm*, and the following statements hold:
(a) $\|x+y\| \leqslant \|x\| + \|y\|$ for all x and y from V;
(b) $\|\alpha x\| = |\alpha|\|x\|$ for any $x \in V$ and scalar α;
(c) $\|x\| \geqslant 0, x \in V$, and $\|x\| = 0$ if and only if $x = 0$.

If a nonnegative function $p(x)$ on a linear space V satisfies the conditions $p(x+y) \leqslant p(x) + p(y)$ and $p(\alpha x) = |\alpha| p(x)$ only, then it is called a *seminorm* on V. A seminorm can return 0 for some nonzero $x \in V$.

Any normed space can be considered as a metric space with the distance $\rho(x,y)$ between x and y given by the formula $\rho(x,y) = \|x-y\|$.

The notation $x = \sum_{n=1}^{\infty} x_n$ means that $\|x - \sum_{n=1}^{N} x_n\| \to 0$ as $N \to \infty$.

Definition 2.23. A *Banach space* is a normed space complete with respect to the metric generated by the norm.

The simplest examples of complex and real Banach spaces are the sets $\mathbb{C}$ and $\mathbb{R}$ themselves.

Example 2.1. Let Q be a bounded domain in $\mathbb{R}^n$. The set $C(\overline{Q})$ of all complex (real) functions continuous in $\overline{Q}$ is a complex (real) Banach space with the norm

$$\|f\|_{C(\overline{Q})} = \max_{x\in\overline{Q}} |f(x)|$$

and standard pointwise operations of summation of functions and multiplication by complex (real) numbers. Convergence in the space $C(\overline{Q})$ is the *uniform* convergence.

Example 2.2. For a nonnegative integer k, the set $C^k(\overline{Q})$ (see Definition 1.6) is a Banach space with respect to the norm

$$\|f\|_{C^k(\overline{Q})} = \max_{|\alpha|\leqslant k} \max_{x\in\overline{Q}} |D^\alpha f(x)|.$$

Example 2.3. Let E be a measurable subset of $\mathbb{R}^n$. The linear space $L_1(E)$ of all Lebesgue-integrable functions $f : E \to \mathbb{C}$ (with the natural identification of functions coinciding almost everywhere) is a Banach space with the norm

$$\|f\|_{L_1(E)} = \int_E |f(x)|\,dx.$$

The convergence $\|f_m - f\|_{L_1(E)} \to 0$, $m \to \infty$, in $L_1(E)$ is called *convergence in the mean.*

Example 2.4. Let $p \geqslant 1$. The space $L_p(E)$ of all measurable functions $f : E \to \mathbb{C}$ such that $|f|^p \in L_1(E)$ is a Banach space with respect to the norm

$$\|f\|_{L_p(E)} = \left(\int_E |f(x)|^p\,dx\right)^{1/p}.$$

Remark 2.6. Let us mention another common way of introducing a metric on a linear space V. Suppose that a sequence of seminorms $p_1, p_2, \ldots$ is given on V, and the condition $p_j(x) = 0$ ($j = 1, 2, \ldots$) implies $x = 0$. It is easy to see then that the function

$$d(x,y) = \sum_{j=0}^{\infty} 2^{-j} \frac{p_j(x-y)}{1 + p_j(x-y)}$$

is a metric on V. The convergence of a sequence x_m to an element x in the metric space V is equivalent to the fact that $p_j(x_m - x) \to 0$ as $m \to \infty$ for all $j = 1, 2, \ldots$. In the described situation, the space V is called *countably normed.*

Definition 2.24. A linear operator $A: X \to Y$ from a normed space X to a normed space Y is called *bounded* if there exists a constant $C > 0$ such that $\|Ax\| \leqslant C\|x\|$ for all $x \in X$. The norm of an operator A is given by the formula

$$\|A\| = \sup\{\|Ax\| : x \in X, \|x\| \leqslant 1\}. \tag{2.1}$$

A linear operator acting in normed spaces is continuous if and only if it is bounded. If $A : X \to Y$ and $B : Y \to Z$ are bounded linear operators in normed spaces, then their composition $BA : X \to Z$ is also a bounded linear operator and $\|BA\| \leqslant \|B\|\|A\|$.

The set of all bounded operators from a Banach space X to a Banach space Y is itself a Banach space with respect to norm (2.1) and the standard pointwise operations of summation of linear operators and their multiplication by scalars.

Definition 2.25. A sequence x_n in a Banach space X *weakly converges* to $x \in X$ if a sequence of numbers $f(x_n)$ converges to $f(x)$ for any bounded linear functional f on X. The notation $x_n \rightharpoonup x, n \to \infty$ is used for weak convergence.

In contrast to weak convergence, a sequence x_n in a Banach space X is called strongly convergent to $x \in X$ if $\|x_n - x\| \to 0$ as $n \to \infty$.

The following statement is known as *Banach's inverse operator theorem.*

Theorem 2.5. *If $A : X \to Y$ is a bounded linear operator mapping a Banach space X one-to-one onto a whole Banach space Y, then the inverse operator $A^{-1} : Y \to X$ is also bounded.*

Theorem 2.6. *Let X be a Banach space, I be the identity operator on X and A be a bounded linear operator $A : X \to X$ such that $\|A\| < 1$. Then there exists a bounded inverse operator $(I - A)^{-1} : X \to X$. This operator can be represented as a series*

$$(I - A)^{-1} = I + A + A^2 + \cdots$$

converging with respect to the operator norm, and $\|(I - A)^{-1}\| \leqslant (1 - \|A\|)^{-1}$.

Definition 2.26. A linear operator from a Banach space X to a Banach space Y is called *compact* if the image of a ball in the space X has a compact closure in Y. Every compact linear operator is bounded, but not vice versa. In the case $\dim Y < \infty$, the boundedness of a linear operator is equivalent to its compactness.

Definition 2.27. A complex linear space H is called a *unitary space* if each ordered pair of vectors $x, y \in H$ is associated with a complex number (x, y) called an *inner product* of x and y, and the following rules are fulfilled:
(a) $(y, x) = \overline{(x, y)}$ (the bar means complex conjugation).
(b) $(x + y, z) = (x, z) + (y, z)$ for all $x, y, z \in H$.
(c) $(\alpha x, y) = \alpha(x, y)$ for $x, y \in H$ and a scalar α.
(d) $(x, x) \geqslant 0$ for all $x \in H$.
(e) $(x, x) = 0$ if and only if $x = 0$.

One can introduce a norm in a unitary space putting $\|x\| = \sqrt{(x, x)}$. Thus, every unitary space is normed and, thereby, a metric space.

Definition 2.28. A *Hilbert space* is an infinite-dimensional unitary space complete with respect to the metric generated by the norm.

Any inner product satisfies the *Cauchy–Schwarz inequality*

$$|(x,y)| \leqslant \|x\|\|y\| \quad (x,y \in H).$$

The following useful statement is known as the *Riesz theorem.*

Theorem 2.7. *For every bounded linear functional f in a Hilbert space H, there exists a unique vector F from H such that*

$$f(x) = (x,F) \quad (x \in H),$$

where $\|F\| = \|f\|$.

Definition 2.29. For any bounded linear operator T in a Hilbert space H, the relation

$$(Tx,y) = (x,T^*y) \quad (x,y \in H)$$

uniquely defines a bounded linear operator T^* in H, called *adjoint* to T (in the sense of a Hilbert space).

There are obvious equalities $\|T^*\| = \|T\|$, $(T+S)^* = T^* + S^*$, $(\alpha T)^* = \bar{\alpha}T^*$, $(ST)^* = T^*S^*$ and $T^{**} = T$.

The operator T^* is compact if and only if the operator T is compact.

Theorem 2.8. *The closed ball $\{x \in H : \|x\| \leqslant 1\}$ of a Hilbert space H is weakly compact, i. e., every sequence of vectors x_n, $\|x_n\| \leqslant 1$ contains a subsequence weakly convergent to some vector $x \in H$, $\|x\| \leqslant 1$.*

Vectors $x, y \in H$ *are orthogonal* if $(x,y) = 0$. Fixed a sequence e_n of pairwise orthogonal unit vectors of a Hilbert space H, we can associate each vector $x \in H$ with the sequence $\hat{x}_n = (x, e_n)$ of its *Fourier coefficients*. The Fourier coefficients satisfy the *Bessel inequality*

$$\sum_{n=1}^{\infty} |\hat{x}_n|^2 \leqslant \|x\|^2.$$

The series $\sum_{n=1}^{\infty} \hat{x}_n e_n$ converges in H, and the difference $x - \sum_{n=1}^{\infty} \hat{x}_n e_n$ is orthogonal to all vectors e_n, $n = 1, 2, \ldots$.

Theorem 2.9. *For an orthonormal system e_n, $n = 1, 2, \ldots$, in a Hilbert space H, the following three conditions are equivalent:*

(a) *Finite linear combinations of vectors $e_1, e_2, \ldots$ are dense in H.*

(b) $x = \sum_{n=1}^{\infty} \hat{x}_n e_n$ *for all $x \in H$.*

(c) $(x,y) = \sum_{n=1}^{\infty} \hat{x}_n \bar{\hat{y}}_n$ *for all $x, y \in H$.*

By setting $y = x$ in relation (c), we obtain *Parseval's identity*

$$\|x\|^2 = \sum_{n=1}^{\infty} |\hat{x}_n|^2 \quad (x \in H).$$

A countable system $\{e_n\}$ in the context of the theorem is called *an orthonormal basis* of the Hilbert space H.

Theorem 2.10. *In any separable Hilbert space H, there exists an orthonormal basis. If a countable linear independent system of vectors $g_1, g_2, \dots$ is such that the linear manifold spanned by this system is everywhere dense in H, then an orthonormal basis in H can be obtained by applying the Gram–Schmidt orthogonalization process*

$$e_1 = \frac{g_1}{\|g_1\|}, \quad e_2 = \frac{g_2 - (g_2, e_1)e_1}{\|g_2 - (g_2, e_1)e_1\|}, \quad \dots$$
$$\dots, \quad e_n = \frac{g_n - (g_n, e_1)e_1 - (g_n, e_2)e_2 - \dots - (g_n, e_{n-1})e_{n-1}}{\|g_n - (g_n, e_1)e_1 - (g_n, e_2)e_2 - \dots - (g_n, e_{n-1})e_{n-1}\|}, \quad \dots$$

Example 2.5. The space $L_2(Q)$ is a Hilbert space with the inner product

$$(u, v)_{L_2(Q)} = \int_Q u(x)\overline{v}(x)\,dx.$$

If $|Q| < \infty$, then the Cauchy–Schwarz inequality gives $\|u\|_{L_1(Q)} \leqslant |Q|^{1/2}\|u\|_{L_2(Q)}$, i. e., $L_2(Q)$ is continuously embedded in $L_1(Q)$.

Bochner integral

In a number of problems, it is necessary to extend the concept of an integral to functions $f : E \to B$ taking values not in $\mathbb{R}$ or $\mathbb{C}$ but in some Banach space B. Let us introduce a corresponding construction called the *Bochner integral*, following monograph [4, Chapter V, Section 5]. We need such integral in Chapter 5. As before, we consider the case of Lebesgue measure μ in $\mathbb{R}^n$ and a measurable set E in $\mathbb{R}^n$.

A function $s : E \to B$ is called *simple* if it takes constant values $c_i \in B$, $c_i \neq 0$ $(i = 1, \dots, N)$, on a finite set of disjoint measurable sets $E_i \subset E$ such that $\mu(E_i) < \infty$, and $s(x) = 0$ on $E \setminus \bigcup_{i=1}^N E_i$. An integral $\int_E s(x)\,dx$ of a simple function s over the set E is, by definition, the sum $\sum c_i\mu(E_i) \in B$.

A function $f : E \to B$ is said to be *strongly measurable* if there exists a sequence s_k of simple functions strongly converging to f almost everywhere on E, i. e., $\|f(x) - s_k(x)\| \to 0$ for almost all $x \in E$. All numerical functions $\|f(x) - s_k(x)\|$ turn out to be measurable, $\|.\| = \|.\|_B$.

A strongly-measurable function $f : E \to B$ is said to be *Bochner integrable* if there exists a sequence s_k of simple functions strongly converging to f almost everywhere on E such that

$$\lim_{k\to\infty} \int_E \|f(x) - s_k(x)\|\, dx = 0.$$

The latter relation guarantees strong convergence of the sequence $\int_E s_k(x)dx$ in B. The limit

$$\lim_{k\to\infty} \int_E s_k(x)\, dx$$

is called the *Bochner integral* of the function f over the set E and is denoted by $\int_E f(x)\, dx$. Note that this limit does not depend on the choice of an approximating sequence.

Now we introduce a key property of the Bochner integral, which follows from its definition and the Lebesgue theorem on bounded convergence.

A strongly-measurable function $f : E \to B$ is Bochner integrable if and only if the numerical function $\|f(x)\|$ is integrable.

Based on this property, one can prove the following statements:

(a) *any function $f : E \to B$ continuous on a compact set E is Bochner integrable*;

(b) *the estimate*

$$\left\| \int_E f(x)\, dx \right\| \leqslant \int_E \|f(x)\|\, dx \tag{2.2}$$

holds for any Bochner integrable function f;

(c) *if $T : B_1 \to B_2$ is a bounded linear operator from a Banach space B_1 to a Banach space B_2 and $f(x)$ is a Bochner integrable B_1-valued function on E, then the B_2-valued function $Tf(x)$ is also Bochner integrable on E, and*

$$\int_E Tf(x)\, dx = T \int_E f(x)\, dx.$$

Fourier transform

This section is devoted to the Fourier transform of functions of multiple variables, with emphasis on the Plancherel theorem. For the most part, only sketches of proofs are given; see [15, Chapter 7] for more details. The important role that the Fourier transform plays in the theory of partial differential equations is that it allows one to reduce problems of analysis to algebraic problems, i. e., differentiation and convolution turn to the algebraic operation of multiplication.

Definition 2.30. The *Fourier transform* (or the *Fourier image*) of a function $f \in L_1(\mathbb{R}^n)$ is the function

$$\hat{f}(\xi) = (2\pi)^{-n/2} \int_{\mathbb{R}^n} f(x) e^{-i(\xi,x)}\, dx \quad (\xi \in \mathbb{R}^n), \tag{2.3}$$

where i denotes the imaginary unit and $(\xi, x) = \xi_1 x_1 + \cdots + \xi_n x_n$. The phrase *Fourier transform* is also used when talking about the mapping $f(x) \mapsto \hat{f}(\xi)$.

Obviously, the function $\hat{f}(\xi)$ is bounded in $\mathbb{R}^n$, and

$$|\hat{f}(\xi)| \leqslant (2\pi)^{-n/2} \|f\|_{L_1(\mathbb{R}^n)}.$$

The last inequality means that the convergence of a sequence of functions in $L_1(\mathbb{R}^n)$ implies the uniform convergence of the sequence of their Fourier images.

The properties of the Fourier transform given below are rather elementary. If $f(x) \mapsto \hat{f}(\xi)$ and $g(x) \mapsto \hat{g}(\xi)$, then

- $f(x-a) \mapsto e^{-i(a,\xi)}\hat{f}(\xi)$, $a \in \mathbb{R}^n$;
- $f(x/\lambda) \mapsto \lambda^n \hat{f}(\lambda\xi)$, $\lambda > 0$;
- $e^{i(a,x)} f(x) \mapsto \hat{f}(\xi - a)$, $a \in \mathbb{R}^n$;
- $(f * g)(x) \mapsto (2\pi)^{n/2}\hat{f}(\xi)\hat{g}(\xi)$.

Let us comment on the last property. If functions f and g belong to $L_1(\mathbb{R}^n)$, then their convolution

$$(f * g)(x) = \int_{\mathbb{R}^n} f(x-y) g(y)\, dy$$

is defined for almost all $x \in \mathbb{R}^n$ and also belongs to $L_1(\mathbb{R}^n)$. The Fourier transform turns the convolution into the algebraic product $(2\pi)^{n/2}\hat{f}(\xi)\hat{g}(\xi)$. This follows from Fubini's theorem.

In the context of the Fourier transform, the *Schwartz space of rapidly decreasing functions* is usually considered.

Definition 2.31. A function $f \in C^\infty(\mathbb{R}^n)$ is called rapidly decreasing if

$$p_N(f) = \sup_{|\alpha| \leqslant N} \sup_{x \in \mathbb{R}^n} (1 + |x|^2)^N |D^\alpha f(x)| < \infty \quad (N = 0, 1, \ldots). \tag{2.4}$$

In other words, a function decreases at infinity faster than any (negative) power of $|x|$, and the same applies to all of its derivatives. Such functions form a linear space denoted by $\mathcal{S}(\mathbb{R}^n)$.

Since smooth compactly supported functions are dense everywhere in $L_1(\mathbb{R}^n)$ (see Exercise 5), and $C_0^\infty(\mathbb{R}^n) \subset \mathcal{S}(\mathbb{R}^n) \subset L_1(\mathbb{R}^n)$, the space $\mathcal{S}(\mathbb{R}^n)$ is also dense everywhere in

$L_1(\mathbb{R}^n)$. Seminorms (2.4) turn $\mathcal{S}(\mathbb{R}^n)$ into a countably-normed space (Remark 2.6). Convergence in $\mathcal{S}(\mathbb{R}^n)$ is equivalent to the uniform convergence of sequences $x^\beta D^\alpha f_m$ for all multiindices α and β where, as usual, $x^\beta = x_1^{\beta_1} \dots x_n^{\beta_n}$. The metric space $\mathcal{S}(\mathbb{R}^n)$ is complete.

Differentiation and multiplication by a rapidly decreasing function or by an infinitely smooth function growing no faster than a polynomial along with all its derivatives are obviously continuous linear operators in $\mathcal{S}(\mathbb{R}^n)$.

Consider the Fourier image of a rapidly decreasing function. For $f \in \mathcal{S}(\mathbb{R}^n)$, one can differentiate as many times as desired under the integral sign in (2.3), obtaining

$$x^\beta f(x) \mapsto i^{|\beta|} D^\beta \widehat{f}(\xi) \tag{2.5}$$

for any multiindex β. On the other hand, repeated integration by parts gives

$$D^\alpha f(x) \mapsto i^{|\alpha|} \xi^\alpha \widehat{f}(\xi) \tag{2.6}$$

for all multiindexes α. Combining (2.5) and (2.6), one can see that the function $i^{|\alpha|+|\beta|} \xi^\alpha D^\beta \widehat{f}(\xi)$ is the Fourier image of the function $D^\alpha(x^\beta f(x))$. This means, in particular, that the Fourier transform maps rapidly decreasing functions to rapidly decreasing functions, i. e., it acts as a linear operator in the space $\mathcal{S}(\mathbb{R}^n)$. This operator is continuous. Moreover, the following statement holds.

Theorem 2.11. *The Fourier transform is a continuous linear one-to-one mapping of the space $\mathcal{S}(\mathbb{R}^n)$ onto itself, and the inverse mapping is also continuous.*

Corollary 2.1. *The Fourier image $\widehat{f}(\xi)$ of any integrable function $f(x)$ is a function continuous in $\mathbb{R}^n$ and $\widehat{f}(\xi) \to 0$ as $|\xi| \to \infty$.*

It suffices to note that $\widehat{f}$ is the uniform limit of a sequence of rapidly decreasing functions.

Our goal now is to obtain a formula for inverting the Fourier transform. The function $\varphi(x) = e^{-|x|^2/2}$ plays an important role here. Let us obtain $\widehat{\varphi}$. Since

$$\widehat{\varphi}(\xi) = \prod_{j=1}^{n} \frac{1}{\sqrt{2\pi}} \int_{-\infty}^{+\infty} e^{-x_j^2/2 - i\xi_j x_j}\, dx_j,$$

the calculation comes down to the case $n = 1$, $\varphi(x) = e^{-x^2/2}$, $x \in \mathbb{R}$. But then the function φ satisfies the ordinary differential equation

$$\varphi'(x) + x\varphi(x) = 0 \quad (x \in \mathbb{R}), \tag{2.7}$$

and its Fourier image $\widehat{\varphi}(\xi)$ satisfies the equation

$$i\xi\widehat{\varphi}(\xi) + i\widehat{\varphi}'(\xi) = 0 \quad (\xi \in \mathbb{R})$$

coinciding with (2.7). Hence $\widehat{\varphi}(\xi)$ is proportional to $\varphi(\xi)$, i. e., $\widehat{\varphi}(\xi) = c\varphi(\xi)$, $\xi \in \mathbb{R}$. We obtain the constant by comparing the right- and the left-hand sides of this equality for $\xi = 0$:

$$c = \frac{1}{\sqrt{2\pi}} \int_{-\infty}^{+\infty} e^{-x^2/2}\, dx = \frac{1}{\sqrt{\pi}} \int_{-\infty}^{+\infty} e^{-u^2}\, du = 1.$$

Therefore, $\widehat{\varphi}(\xi) = \varphi(\xi)$, and the same conclusion is valid in $\mathbb{R}^n$.

Applying Fubini's theorem to the double integral

$$\int_{\mathbb{R}^n}\int_{\mathbb{R}^n} f(x)g(y)e^{-i(x,y)}\, dxdy$$

in the case of $f \in L_1(\mathbb{R}^n)$ and $g \in L_1(\mathbb{R}^n)$, we obtain

$$\int_{\mathbb{R}^n} \widehat{f}(y)g(y)\, dy = \int_{\mathbb{R}^n} f(x)\widehat{g}(x)\, dx. \tag{2.8}$$

Set $g \in \mathcal{S}(\mathbb{R}^n)$ and $f(x) = \varphi(x/\lambda)$, $\lambda > 0$. Then

$$\int_{\mathbb{R}^n} \widehat{\varphi}(x)g(x/\lambda)\, dx = \int_{\mathbb{R}^n} \varphi(x/\lambda)\widehat{g}(x)\, dx.$$

Passing to the limit at $\lambda \to +\infty$ in the last equality and using the Lebesgue theorem, we obtain

$$g(0)\int_{\mathbb{R}^n} \widehat{\varphi}(x)\, dx = \varphi(0)\int_{\mathbb{R}^n} \widehat{g}(x)\, dx.$$

Taking into account that

$$\varphi(0) = 1, \qquad \int_{\mathbb{R}^n} \widehat{\varphi}(x)\, dx = \int_{\mathbb{R}^n} e^{-|x|^2/2}\, dx = (2\pi)^{n/2},$$

we write

$$g(0) = (2\pi)^{-n/2}\int_{\mathbb{R}^n} \widehat{g}(\xi)\, d\xi.$$

Substituting $g(x+y)$ for $g(x)$ in the resulting relation, we arrive at the formula

$$g(y) = (2\pi)^{-n/2}\int_{\mathbb{R}^n} \widehat{g}(\xi)e^{i(y,\xi)}\, d\xi \tag{2.9}$$

that restores a rapidly decreasing function from its Fourier image.

Definition 2.32. The right-hand side in (2.9) is called the *inverse Fourier transform.*

Formula (2.9) can easily be extended to the case of an integrable function f whose Fourier image $\widehat{f}$ is also integrable over $\mathbb{R}^n$. Let g belong to $\mathcal{S}(\mathbb{R}^n)$ as before. Then the integral $(2\pi)^{-n/2}\int_{\mathbb{R}^n}\widehat{g}(x)e^{i(y,x)}\,dx$ can replace $g(y)$ in the left-hand side of equality (2.8). By Fubini's theorem, we have

$$(2\pi)^{-n/2}\int\limits_{\mathbb{R}^n}\widehat{g}(x)\,dx\int\limits_{\mathbb{R}^n}\widehat{f}(y)e^{i(y,x)}\,dy=\int\limits_{\mathbb{R}^n}\widehat{g}(x)f(x)\,dx.$$

Since $\widehat{g}$ runs over the entire space $\mathcal{S}(\mathbb{R}^n)$, the equality

$$f(x)=(2\pi)^{-n/2}\int\limits_{\mathbb{R}^n}\widehat{f}(y)e^{i(y,x)}\,dy$$

holds for almost all $x\in\mathbb{R}^n$ (see Exercise 4).

Now everything is set to prove an extremely important result called the *Plancherel theorem* (Theorem 2.12 below).

For $f\in\mathcal{S}(\mathbb{R}^n)$ and $g\in\mathcal{S}(\mathbb{R}^n)$, using inversion formula (2.9) and Fubini's theorem, we can write

$$\begin{aligned}\int\limits_{\mathbb{R}^n}f(x)\overline{g(x)}\,dx&=(2\pi)^{-n/2}\int\limits_{\mathbb{R}^n}\overline{g(x)}\,dx\int\limits_{\mathbb{R}^n}\widehat{f}(\xi)e^{i(\xi,x)}\,d\xi=\\&=(2\pi)^{-n/2}\int\limits_{\mathbb{R}^n}\widehat{f}(\xi)\,d\xi\int\limits_{\mathbb{R}^n}\overline{g(x)}e^{i(\xi,x)}\,dx=\\&=(2\pi)^{-n/2}\int\limits_{\mathbb{R}^n}\widehat{f}(\xi)\,d\xi\overline{\int\limits_{\mathbb{R}^n}g(x)e^{-i(\xi,x)}\,dx}=\int\limits_{\mathbb{R}^n}\widehat{f}(\xi)\overline{\widehat{g}(\xi)}\,d\xi,\end{aligned}$$

i. e.,

$$(f,g)_{L_2(\mathbb{R}^n)}=(\widehat{f},\widehat{g})_{L_2(\mathbb{R}^n)},\quad\|f\|_{L_2(\mathbb{R}^n)}=\|\widehat{f}\|_{L_2(\mathbb{R}^n)}.\tag{2.10}$$

This shows that the Fourier transform is an isometry of the linear space $\mathcal{S}(\mathbb{R}^n)$ everywhere dense in $L_2(\mathbb{R}^n)$ onto itself. But then it can be uniquely extended to the isometry of the entire $L_2(\mathbb{R}^n)$. This extension is called the *Fourier–Plancherel transform.* However, the former term *Fourier transform* is usually used for it and the previous notation $\widehat{f}$ is retained. Taking above into account, formulas (2.10) extend to arbitrary functions f and g from $L_2(\mathbb{R}^n)$.

Theorem 2.12. *The Fourier transform is an isometric isomorphism of the Hilbert space $L_2(\mathbb{R}^n)$ onto itself.*

Definition 2.33. Formulas (2.10) are called *Parseval's identities.*

Remark 2.7. Let a function f from $L_2(\mathbb{R}^n)$ be such that $f(x) = 0$ for $|x| \geqslant R > 0$. Then f belongs to $L_1(\mathbb{R}^n)$, and its Fourier transform $\widehat{f}(\xi)$ is defined in the form of integral (2.3), which is a continuous function vanishing at infinity. Let us approximate the function f in the space $L_2(\mathbb{R}^n)$ by a sequence of functions $f_k \in \mathcal{S}(\mathbb{R}^n)$ supported in the ball $|x| \leqslant R$. Such a sequence f_k converges to f in the norm of the space $L_1(\mathbb{R}^n)$ as well. Therefore, the sequence of Fourier transforms $\widehat{f}_k(\xi)$ converges to $\widehat{f}(\xi)$ uniformly on $\mathbb{R}^n$. Moreover, $\widehat{f}_k$ converges in $L_2(\mathbb{R}^n)$ to the Fourier–Plancherel transform of f by Theorem 2.12. Thus, we are now convinced that the Fourier–Plancherel transform of any compactly supported function from $L_2(\mathbb{R}^n)$ is still given by integral (2.3).

Now let f be an arbitrary function from $L_2(\mathbb{R}^n)$. Consider the family of truncated functions ($R > 0$)

$$f_R(x) = \begin{cases} f(x), & |x| < R, \\ 0, & |x| \geqslant R, \end{cases}$$

for which

$$\widehat{f}_R(\xi) = (2\pi)^{-n/2} \int\limits_{|x|<R} f(x) e^{-i(\xi,x)}\, dx.$$

Obviously, $f_R \to f$ in $L_2(\mathbb{R}^n)$ as $R \to +\infty$. Therefore, by Theorem 2.12, the Fourier–Plancherel image of an arbitrary function $f \in L_2(\mathbb{R}^n)$ can be defined as the limit

$$\widehat{f}(\xi) = \lim_{R\to+\infty} (2\pi)^{-n/2} \int\limits_{|x|<R} f(x) e^{-i(\xi,x)}\, dx$$

understood in the mean-square sense, i. e., in $L_2(\mathbb{R}^n)$.

Fredholm theory

The results of this and the next sections are thoroughly used in the main part of the book. Therefore, for the sake of completeness, full proofs are given. Fredholm's theorems emerged originally in the study of integral equations are presented here in abstract form.

Let H be an infinite-dimensional separable complex Hilbert space and $A : H \to H$ be a compact linear operator. Recall that the latter means that A maps bounded sets to precompact sets in H. In other words, for any bounded sequence $\{x_n\}_{n=1}^{\infty} \subset H$, the sequence $\{Ax_n\}_{n=1}^{\infty}$ contains a Cauchy subsequence. Note that a compact linear operator takes any weakly-convergent sequence to the strongly-convergent one.

The canonical Fredholm operator has the form

$$T = I - A : H \to H.$$

Like for any bounded linear operator, its kernel (null space)

$$\mathcal{N}(T) = \{x \in H : Tx = 0\}$$

is a closed linear subspace in H.

Lemma 2.1. *The image $\mathcal{R}(T) = T(H) = \{Tx : x \in H\}$ of the operator T is a closed linear subspace in H.*

Proof. Let $y_n \in \mathcal{R}(T)$ and $y_n \to y$ in H. This means the existence of vectors $x_n \in H$ such that

$$y_n = Tx_n = x_n - Ax_n \to y.$$

Without loss of generality, we can assume that x_n are orthogonal to $\mathcal{N}(T)$; otherwise, we subtract from the vector x_n its projection onto $\mathcal{N}(T)$, thus the value of Tx_n does not change. Let us show that the sequence x_n is then bounded in H. Assuming the contrary, we move on to an infinitely large subsequence, for which we retain the previous notation x_n, $\|x_n\| \to \infty$, and introduce the sequence $z_n = x_n/\|x_n\|$ lying on the unit sphere in the orthogonal complement of $\mathcal{N}(T)$. We have

$$z_n - Az_n \to 0. \tag{2.11}$$

Passing to a subsequence again, we can assume that Az_n converges in H due to compactness of the operator A. It follows from (2.11) that the sequence z_n converges itself to some vector $z \in H$ such that $\|z\| = 1$ and $Tz = 0$, i. e., $z \in \mathcal{N}(T)$. On the other hand, the vector z is orthogonal to $\mathcal{N}(T)$ together with z_n.

The obtained contradiction proves the boundedness of the sequence x_n. Then some subsequence of the sequence Ax_n converges in H, and together with the convergence of $x_n - Ax_n \to y$ this means the existence of a convergent subsequence of x_n. For its limit $x \in H$, we have $y = Tx$, i. e., $y \in \mathcal{R}(T)$. □

In Fredholm theory, the operator T is considered together with its adjoint operator $T^* = I - A^*$. The image of $\mathcal{R}(T^*)$ is also closed in H due to the compactness of A^*.

Now it is easy to obtain decompositions

$$H = \mathcal{N}(T) \oplus \mathcal{R}(T^*) = \mathcal{N}(T^*) \oplus \mathcal{R}(T). \tag{2.12}$$

Indeed, if $h \in \mathcal{N}(T)$, then $(h, T^*x) = (Th, x) = 0$ for all $x \in H$, i. e., h is orthogonal to $\mathcal{R}(T^*)$. On the other hand, let a vector $z \in H$ be orthogonal to $\mathcal{R}(T^*)$. Then we have $(Tz, x) = (z, T^*x) = 0$ for all $x \in H$, i. e., $Tz = 0$. Taking into account that $\mathcal{N}(T)$ and $\mathcal{R}(T^*)$ are closed subspaces of H, we arrive at the equality $H = \mathcal{N}(T) \oplus \mathcal{R}(T^*)$. The second equality is checked in a similar way.

The first of Fredholm's three theorems is obtained. It is usually formulated as follows.

Theorem 2.13. *The equation $T\varphi = f$ is solvable if and only if $(f, \psi) = 0$ for any solution of the homogeneous adjoint equation $T^*\psi = 0$.*

Further we consider a sequence of nested subspaces

$$H \supset H^1 \supset H^2 \supset \dots,$$

where $H^1 = T(H)$ and $H^{k+1} = T(H^k)$, $k = 1, 2, \dots$. They are all closed in H by Lemma 2.1.

Lemma 2.2. *There exists a natural number j such that $H^{j+1} = H^j$.*

Proof. If the stated number j does not exist, then all closed subspaces H^k are distinct. But then it is possible to construct an orthonormal sequence $x_k \in H^k$ such that the vector x_k is orthogonal to the subspace H^{k+1}. Take $l > k$ and consider the difference

$$Ax_l - Ax_k = -x_k + (x_l + Tx_k - Tx_l).$$

Since the vector in brackets belongs to the space H^{k+1}, we have

$$\|Ax_l - Ax_k\|^2 = \|x_k\|^2 + \|x_l + Tx_k - Tx_l\|^2 \geqslant 1,$$

whence it follows that the sequence Ax_k has no convergent subsequences. The latter contradicts the compactness of the operator A. □

Now we see that $\mathcal{N}(T) = \{0\}$ implies $\mathcal{R}(T) = H$. Indeed, if the null space of T is trivial and $\mathcal{R}(T) \neq H$, then $H^{k+1} \neq H^k$ for all $k = 1, 2, \dots$ since the mapping T is one-to-one. This contradicts Lemma 2.2. Similarly, it follows from $\mathcal{N}(T^*) = \{0\}$ that $\mathcal{R}(T^*) = H$.

Conversely, let $\mathcal{R}(T) = H$. Then $\mathcal{N}(T^*) = \{0\}$ due to relation (2.12), thus $\mathcal{R}(T^*) = H$. Applying (2.12) again, we conclude that $\mathcal{N}(T) = \{0\}$.

Therefore, the *Fredholm alternative* is proven.

Theorem 2.14. *Either the equation $T\varphi = f$ has a unique solution for any vector $f \in H$, or the homogeneous equation $T\varphi = 0$ has a nontrivial solution.*

It is easy to see that $\dim \mathcal{N}(T) < \infty$, and the same is true for the operator T^*. Indeed, if the subspace $\mathcal{N}(T)$ is infinite-dimensional, then there exists an orthonormal sequence $\{x_n\}_{n=1}^{\infty}$ for which $Tx_n = 0$, i. e., $Ax_n = x_n$. It follows that $\|Ax_k - Ax_l\| = \sqrt{2}$ for $k \neq l$, and this contradicts the compactness of the operator A.

Let us denote $\mu = \dim \mathcal{N}(T)$ and $\nu = \dim \mathcal{N}(T^*)$. It remains to make sure that $\mu = \nu$. Assume the opposite, e. g., $\mu < \nu$. Let $\{\varphi_1, \dots, \varphi_\mu\}$ be an orthonormal basis in $\mathcal{N}(T)$ and $\{\psi_1, \dots, \psi_\nu\}$ be an orthonormal basis in $\mathcal{N}(T^*)$. Introduce an operator S by the formula

$$Sx = Tx + \sum_{j=1}^{\mu} (x, \varphi_j)\psi_j.$$

Let us show that the equation $Sx = 0$ has the only trivial solution. In the equality

$$Tx + \sum_{j=1}^{\mu}(x, \varphi_j)\psi_j = 0, \tag{2.13}$$

the vector Tx is orthogonal to the vectors ψ_j due to relations (2.12). Therefore, (2.13) implies $Tx = 0$ and $(x, \varphi_j) = 0$ $(j = 1, \dots, \mu)$. So, on the one hand, a solution x to equation (2.13) is a linear combination of the vectors $\varphi_1, \dots, \varphi_\mu$, and on the other hand, it is orthogonal to these vectors, i. e., $x = 0$.

Since S differs from T by a finite-dimensional (and, therefore, compact) operator, all the results obtained above, including Theorem 2.14, are also valid for S. Therefore, there exists a vector $y \in H$ such that

$$Ty + \sum_{j=1}^{\mu}(y, \varphi_j)\psi_j = \psi_{\mu+1}.$$

However, the vectors $\psi_{\mu+1}$ and $Ty + \sum_{j=1}^{\mu}(y, \varphi_j)\psi_j$ are orthogonal, which leads to a contradiction.

Therefore, the following statement has been proven.

Theorem 2.15. *The homogeneous equations $T\varphi = 0$ and $T^*\psi = 0$ have the same finite number of linearly independent solutions.*

At the end of this section, we consider the issue of (nonzero) eigenvalues of a compact operator A. Let us denote $T_\lambda = \lambda I - A$, $0 \neq \lambda \in \mathbb{C}$, then $T_\lambda^* = \bar{\lambda} I - A^*$. Recall that a number $\lambda \in \mathbb{C}$ is an eigenvalue of the operator A if $\mathcal{N}(T_\lambda) \neq \{0\}$ (in this case, nonzero elements of $\mathcal{N}(T_\lambda)$ are called eigenvectors and the subspace $\mathcal{N}(T_\lambda)$ itself the eigenspace). All the above statements concerning the operators T and T^* obviously carry over to the operators T_λ and T_λ^*, $0 \neq \lambda \in \mathbb{C}$. In particular, all nonzero eigenvalues of a compact operator have finite multiplicity (the corresponding eigenspaces are finite-dimensional). A number λ is an eigenvalue of the operator A if and only if $\bar{\lambda}$ is an eigenvalue of the operator A^*.

Lemma 2.3. *Outside a circle of arbitrary radius $r > 0$ centered at zero there exists only a finite number of eigenvalues of a compact operator.*

Proof. Assume the opposite: $\{\lambda_n\}_{n=1}^{\infty}$ is a sequence of distinct eigenvalues of a compact operator A with $|\lambda_n| \geqslant r$. By choosing one eigenvector e_n for each λ_n, we obtain a linearly independent system $e_1, e_2, \dots$. We denote the linear span of the first n elements of this system by M_n. We have strict embeddings $M_1 \subset M_2 \subset \dots$, and $T_{\lambda_n}(M_n) \subset M_{n-1}$. Indeed, if $x = \alpha_1 e_1 + \dots + \alpha_n e_n$, then

$$T_{\lambda_n} x = \alpha_1 \lambda_n e_1 + \dots + \alpha_n \lambda_n e_n - \alpha_1 \lambda_1 e_1 - \dots - \alpha_n \lambda_n e_n \in M_{n-1}.$$

Let $x_n \in M_n$ be an orthonormal sequence such that the vector x_n is orthogonal to the subspace M_{n-1}. For $2 \leqslant m < n$, we have $T_{\lambda_n} x_n + \lambda_m x_m - T_{\lambda_m} x_m \in M_{n-1}$, thus

$$\begin{aligned}\|Ax_n - Ax_m\| &= \|\lambda_n x_n - T_{\lambda_n} x_n - \lambda_m x_m + T_{\lambda_m} x_m\| = \\ &= |\lambda_n| \|x_n - \lambda_n^{-1}(T_{\lambda_n} x_n + \lambda_m x_m - T_{\lambda_m} x_m)\| \geqslant r.\end{aligned}$$

The obtained inequality contradicts the compactness of the operator A. □

Lemma 2.3 means that the set of eigenvalues of a compact operator is at most countable and has no accumulation points except maybe the origin.

The Hilbert–Schmidt theorem

Recall that a bounded linear operator A acting in a Hilbert space H is called self-adjoint if

$$(A\xi, \eta) = (\xi, A\eta) \quad (\xi, \eta \in H).$$

A self-adjoint operator $A : H \to H$ is called positive if $(A\xi, \xi) > 0$ for all $0 \neq \xi \in H$.

One can easily check that all eigenvalues of a self-adjoint operator are real and that all eigenvalues of a positive operator are positive.

Theorem 2.16. *For every compact self-adjoint linear operator A in a separable Hilbert space H, there exists an orthonormal basis $\{\varphi_n\}_{n=1}^{\infty}$ in H consisting of eigenvectors of the operator A.*

Proof. Consider the quadratic form $Q(\xi) = (A\xi, \xi)$ of the operator A. Without loss of generality, we assume that $Q(\xi)$ is not identically zero since $(A\xi, \xi) \equiv 0$ implies $A = 0$ for any bounded linear operator in a complex Hilbert space. Take an arbitrary sequence of points $\xi_1, \xi_2, \dots$ of the unit ball, maximizing $|Q(\xi)|$,

$$|Q(\xi_n)| \to S = \sup_{\|\xi\| \leqslant 1} |Q(\xi)| > 0.$$

Since a ball in H is weakly compact, there exists a subsequence that weakly converges to some vector $\varphi \in H$, $\|\varphi\| \leqslant 1$. Let us keep the same notation ξ_n for this subsequence. The compactness of A implies $A\xi_n \to A\varphi$ with respect to the norm in H. Then we have

$$\begin{aligned}|Q(\xi_n) - Q(\varphi)| &\leqslant |(A\xi_n, \xi_n) - (A\varphi, \xi_n)| + |(A\varphi, \xi_n) - (A\varphi, \varphi)| = \\ &= |(A\xi_n - A\varphi, \xi_n)| + |(A\xi_n - A\varphi, \varphi)| \leqslant \|A\xi_n - A\varphi\|(\|\xi_n\| + \|\varphi\|) \leqslant \\ &\leqslant 2\|A\xi_n - A\varphi\| \to 0, \quad n \to \infty.\end{aligned}$$

We obtain $|Q(\varphi)| = S$, and it is clear that $\|\varphi\| = 1$. Otherwise, if $\|\varphi\| < 1$, then we consider the unit vector $\varphi' = \varphi/\|\varphi\|$ for which $|Q(\varphi')| > S$.

Let us verify the following key property of the resulting vector φ:

$$(\varphi, \eta) = 0 \Longrightarrow (A\varphi, \eta) = 0. \tag{2.14}$$

Suppose that there exists a vector η for which $(\varphi, \eta) = 0$, but $\gamma = (A\varphi, \eta) \neq 0$. Let $\|\eta\| = 1$. Consider the family of vectors

$$\xi = \xi(t) = \frac{\varphi + te^{i \arg \gamma}\eta}{\sqrt{1 + t^2}}, \quad t \in \mathbb{R},$$

on the unit sphere. We have

$$Q(\xi(t)) = \frac{1}{1 + t^2}(Q(\varphi) + 2t|\gamma| + t^2 Q(\eta))$$

and

$$\left.\frac{dQ(\xi(t))}{dt}\right|_{t=0} = 2|\gamma| > 0.$$

This contradicts the fact that the maximum $|Q(\xi)|$ on the unit sphere is achieved at the vector $\xi = \varphi$. Relation (2.14) has been proven.

Represent the vector $A\varphi$ in the form $A\varphi = \lambda\varphi + \eta$ with a vector η orthogonal to the vector φ. Then relation (2.14) shows that $(A\varphi, \eta) = \|\eta\|^2 = 0$, i. e., $\eta = 0$. Therefore, φ is an eigenvector of the operator A with the eigenvalue λ. We take φ as the first element of the basis, $\varphi_1 = \varphi$. The corresponding eigenvalue $\lambda_1 = \lambda$ is such that $|\lambda_1| = |Q(\varphi)| = S$.

In order to construct the next basis vector φ_2, we restrict the quadratic form $Q(\xi)$ to the closed subspace

$$M_1 = \{\xi \in H : (\xi, \varphi_1) = 0\},$$

which is an invariant subspace of the operator A (indeed, if $(\xi, \varphi_1) = 0$, then $(A\xi, \varphi_1) = (\xi, A\varphi_1) = \lambda_1(\xi, \varphi_1) = 0$). Then φ_2 is a unit vector for which $|Q(\xi)|$ attains its maximum $|\lambda_2| \leqslant |\lambda_1|$ on the unit sphere in M_1. This is an eigenvector of A with the eigenvalue λ_2. Continuing in a similar manner, we are faced with one of the following two situations:

1) there exists a natural number n such that $Q(\xi)$ vanishes on the subspace

$$M_n = \{\xi \in H : (\xi, \varphi_j) = 0, j = 1, \dots, n\}$$

invariant for the operator A.

2) the restriction of $Q(\xi)$ to M_n is not identically zero for any n.

In the first case, M_n is exactly the kernel of the operator A. It consists of all eigenvectors corresponding to the eigenvalue $\lambda = 0$. By completing the orthonormal system $\varphi_1, \dots, \varphi_n$ of eigenvectors corresponding to nonzero eigenvalues $\lambda_1, \dots, \lambda_n$, with an arbitrary orthonormal basis of the subspace M_n, we obtain an orthonormal basis of H consisting of eigenvectors of the operator A.

In the second case, we have an orthonormal sequence of eigenvectors φ_n such that $\lambda_n \neq 0$ for all $n = 1, 2, \dots$ and

$$|\lambda_1| \geqslant |\lambda_2| \geqslant \cdots \geqslant |\lambda_n| \geqslant \ldots .$$

Like any orthonormal sequence, φ_n weakly converges to zero. Then $\|A\varphi_n\| = |\lambda_n| \to 0$ due to compactness of A. Set

$$M = \{\xi \in H : (\xi, \varphi_j) = 0, j = 1, 2, \ldots\} = \bigcap_{n=1}^{\infty} M_n.$$

It is clear that the operator A maps the subspace M to zero. Indeed, if $\xi \in M$ and $\|\xi\| = 1$, then $|(A\xi, \xi)| \leqslant |\lambda_n|$ for all $n = 1, 2, \ldots$, i. e., $(A\xi, \xi) = 0$. But M is invariant under A, thus A vanishes on M. It remains to supplement the constructed system $\varphi_1, \varphi_2, \ldots$ with an orthonormal basis in M. □

3 Cut-off functions and partition of unity

Cut-off functions

Set

$$\omega(t) = \begin{cases} c_n e^{-(1-t^2)^{-1}}, & |t| < 1, \\ 0, & |t| \geqslant 1, \end{cases}$$

where a positive number c_n is chosen so that

$$\int_{\mathbb{R}^n} \omega(|x|)\, dx = 1.$$

Definition 3.1. The function

$$\omega_h(x) = \frac{1}{h^n}\, \omega\left(\frac{|x|}{h}\right) \quad (x \in \mathbb{R}^n;\, h > 0)$$

is called a *mollifier*.

The following properties of mollifiers are obvious:

- $\omega_h \in C_0^\infty(\mathbb{R}^n)$,
- $\omega_h(x) \geqslant 0$ for $x \in \mathbb{R}^n$ and $\omega_h(x) = 0$ for $|x| \geqslant h$,
- $\int_{\mathbb{R}^n} \omega_h(x)\, dx = 1$,
- $|D^\alpha \omega_h(x)| \leqslant M_\alpha h^{-(n+|\alpha|)}$, where $M_\alpha > 0$ does not depend on h and x.

Consider a bounded domain $G \subset \mathbb{R}^n$ and a positive number δ. Let us introduce an open set $G_\delta = \{x \in G : \rho(x, \partial G) > \delta\}$ (nonempty for sufficiently small δ), where $\rho(x, E) = \inf_{y \in E} |x - y|$ denotes the distance from the point $x \in \mathbb{R}^n$ to the set $E \subset \mathbb{R}^n$. Obviously, $\overline{G_\delta} \subset G$.

The formula

$$\zeta_\delta(x) = \int\limits_{G_{3\delta/4}} \omega_{\delta/4}(x-y)\,dy \quad (x \in \mathbb{R}^n)$$

defines a smooth function ζ_δ supported in G with values from the segment $[0,1]$ and such that

$$\zeta_\delta(x) = 1 \quad \text{for } x \in G_\delta \quad \text{and} \quad \zeta_\delta(x) = 0 \quad \text{for } x \notin G_{\delta/2}. \tag{3.1}$$

Note that the ball $B_{\delta/4}(x) = \{y \in \mathbb{R}^n : |y-x| < \delta/4\}$ is entirely contained in the domain of integration if $x \in G_\delta$. On the other hand, $B_{\delta/4}(x) \cap G_{3\delta/4} = \varnothing$ if $x \notin G_{\delta/2}$. Therefore, relations (3.1) follow from the properties of the mollifier ω_h.

Definition 3.2. The function ζ_δ constructed above is called the *cut-off function* for the domain G.

Partition of unity

Theorem 3.1. *Let Q be a bounded domain and $\{U_\alpha\}_{\alpha\in A}$ be a family of open subsets of $\mathbb{R}^n$ such that $\overline{Q} \subset \bigcup_{\alpha\in A} U_\alpha$. Then there exist nonnegative functions $f_j \in C_0^\infty(\mathbb{R}^n)$ $(j = 1,\dots,N)$ such that*

1) *for each j there exists $\alpha_j \in A$ such that* $\operatorname{supp} f_j \subset U_{\alpha_j}$,
2) $\sum_{j=1}^N f_j(x) \leqslant 1$ *for all* $x \in \mathbb{R}^n$,
3) $\sum_{j=1}^N f_j(x) = 1$ *for* $x \in \overline{Q}$.

Definition 3.3. A finite family of functions $\{f_j\}_{j=1}^N$ guaranteed by Theorem 3.1 is called a *partition of unity subordinated to a given open covering* of $\overline{Q}$.

Proof. Fixed an arbitrary point $y \in \overline{Q}$, choose an element U_{α_y} of the covering $\{U_\alpha\}_{\alpha\in A}$ containing this point and construct a function $\varphi_y \in C_0^\infty(U_{\alpha_y})$ such that $0 \leqslant \varphi_y(x) \leqslant 1$ and $\varphi_y(y) = 1$. If

$$G_y = \{x \in \mathbb{R}^n : \varphi_y(x) > 0\},$$

then G_y is an open neighborhood of the point y. By constructing neighborhoods G_y in this manner for all $y \in \overline{Q}$, we obtain a new cover of the closure $\overline{Q}$ by open sets $\{G_y\}_{y\in\overline{Q}}$. Since the set $\overline{Q}$ is compact, we can extract a finite subcover $\{G_{y_j}\}_{j=1}^N$. Let us denote $G = \bigcup_{j=1}^N G_{y_j}$. Then G is an open neighborhood of $\overline{Q}$. Take a cut-off function $g \in C_0^\infty(G)$ such that $g(x) = 1$ in $\overline{Q}$. We claim that the set of functions

$$f_j(x) = \frac{\varphi_{y_j}(x)}{1 - g(x) + \sum_{j=1}^N \varphi_{y_j}(x)} \quad (x \in \mathbb{R}^n;\ j = 1,\dots,N)$$

is the desired partition of unity. To verify this, first note that the numerator and denominator of the fraction contain smooth functions, and the denominator does not vanish anywhere in $\mathbb{R}^n$. Indeed, if all functions φ_{y_j} are equal to zero at a point x, then it follows from the definition of the set G that $x \notin G$ and, therefore, $g(x) = 0$. Obviously, $\operatorname{supp} f_j \subset U_{a_{y_j}}$. Finally,

$$\sum_{j=1}^{N} f_j(x) = \frac{\sum_{j=1}^{N} \varphi_{y_j}(x)}{1 - g(x) + \sum_{j=1}^{N} \varphi_{y_j}(x)} = 1$$

in $\overline{Q}$ since $g(x) = 1$ for $x \in \overline{Q}$. □

Cut-off functions and the partition of unity are powerful tools of analysis that allow many problems to be reduced to the study of local properties of functions and operators.

2 Functional spaces

4 Lebesgue space $L_2(Q)$

Throughout the book, Q denotes a bounded domain in $\mathbb{R}^n$ with boundary ∂Q of at least the C^0 class to guarantee zero n-dimensional Lebesgue measure for ∂Q.

Theorem 4.1. *The set $C(\overline{Q})$ is everywhere dense in $L_2(Q)$.*

Proof. Let us first verify that the set of all simple functions is dense in $L_2(Q)$, where by a *simple function* we mean a linear combination of characteristic functions of measurable subsets of Q. Assume that a function u from $L_2(Q)$ is nonnegative. Then the nonnegative function u^2 belongs to $L_1(Q)$. It follows from the definition of the Lebesgue integral that for any number $\varepsilon > 0$ there exists a step function φ such that $\int_Q |u^2(x) - \varphi(x)|\,dx < \varepsilon$. Without loss of generality, we assume that $\varphi \geqslant 0$. In view of the obvious relations,

$$|u - \sqrt{\varphi}|^2 = |u - \sqrt{\varphi}| \cdot |u - \sqrt{\varphi}| \leqslant |u - \sqrt{\varphi}| \cdot |u + \sqrt{\varphi}| = |u^2 - \varphi|,$$

we get $\int_Q |u - \sqrt{\varphi}|^2\,dx < \varepsilon$.

The case of a sign-alternating function $u \in L_2(Q)$ is reduced to the case of a nonnegative function by the representation $u = u_+ - u_-$, where

$$u_+(x) = \max(u(x), 0) \geqslant 0, \quad u_-(x) = -\min(u(x), 0) \geqslant 0, \quad u_\pm \in L_2(Q).$$

The transition to the complex-valued function $u = \operatorname{Re} u + i \operatorname{Im} u$ is obvious.

It remains to show that the characteristic function

$$\chi_M(x) = \begin{cases} 1, & x \in M, \\ 0, & x \notin M \end{cases}$$

of any Lebesgue measurable set $M \subset Q$ can be approximated in $L_2(Q)$ by continuous functions.

By virtue of Remark 2.1, for any number $\varepsilon > 0$ there exist a closed set $\mathcal{F}_M$ and an open set $\mathcal{G}_M$ such that $\mathcal{F}_M \subset M \subset \mathcal{G}_M \subset Q$ and $|\mathcal{G}_M \setminus \mathcal{F}_M| < \varepsilon$ (the measure of their difference is less than ε).

Introduce the function

$$\varphi(x) = \frac{\rho(x, Q \setminus \mathcal{G}_M)}{\rho(x, Q \setminus \mathcal{G}_M) + \rho(x, \mathcal{F}_M)}.$$

It is easy to check that the function φ is continuous in $\overline{Q}$. Moreover,

$$\begin{aligned} &\varphi(x) = 0 \quad \text{for } x \notin \mathcal{G}_M, \quad \varphi(x) = 1 \quad \text{for } x \in \mathcal{F}_M, \\ &|\chi_M(x) - \varphi(x)| \leqslant 1 \quad \text{for } x \in Q \end{aligned}$$

https://doi.org/10.1515/9783112229637-002

and

$$\chi_M(x) = \varphi(x) \quad \text{for } x \in \mathcal{F}_M \bigcup (Q \setminus \mathcal{G}_M).$$

Therefore,

$$\int_Q |\chi_M(x) - \varphi(x)|^2\, dx = \int_{\mathcal{G}_M \setminus \mathcal{F}_M} |\chi_M(x) - \varphi(x)|^2\, dx < \varepsilon. \qquad \square$$

Theorem 4.2. *Any function $f \in L_2(Q)$ is continuous in the mean square, i. e., for any number $\varepsilon > 0$ there exists a number $\delta > 0$ such that*

$$\|f(\cdot + z) - f(\cdot)\|_{L_2(Q)} < \varepsilon \quad \textit{for all } |z| < \delta, \tag{4.1}$$

where $\|f(\cdot + z) - f(\cdot)\|^2_{L_2(Q)} = \int_Q |f(x+z) - f(x)|^2\, dx$.

Remark 4.1. In (4.1), it is assumed that the function f is extended by zero to $\mathbb{R}^n \setminus Q$.

Proof. Take $f \in L_2(Q)$ and $\varepsilon > 0$. Set

$$F(x) = \begin{cases} f(x), & x \in Q, \\ 0, & x \notin Q. \end{cases}$$

Choose a number $a > 0$ such that $\overline{Q} \subset B_a = \{x \in \mathbb{R}^n : |x| < a\}$. Theorem 4.1 guarantees the existence of a function $\widetilde{F} \in C(\overline{B_{2a}})$ with $\|F - \widetilde{F}\|_{L_2(B_{2a})} < \varepsilon/3$. Using a cut-off function $\zeta \in C_0^\infty(\mathbb{R}^n)$ such that $0 \leqslant \zeta(x) \leqslant 1$, $\zeta(x) = 1$ for $x \in Q$ and $\operatorname{supp} \zeta \subset B_a$, we introduce a continuous function

$$\Phi(x) = \begin{cases} \zeta(x)\widetilde{F}(x), & x \in B_{2a}, \\ 0, & x \notin B_{2a}. \end{cases}$$

Obviously, $|F(x) - \Phi(x)| \leqslant |F(x) - \widetilde{F}(x)|$ and

$$\|F - \Phi\|_{L_2(\mathbb{R}^n)} = \|F - \Phi\|_{L_2(B_{2a})} \leqslant \|F - \widetilde{F}\|_{L_2(B_{2a})} < \varepsilon/3.$$

Being a continuous compactly supported function, Φ is uniformly continuous in $\mathbb{R}^n$, i. e., there exists a positive number δ, $\delta < a$, such that

$$|\Phi(x+z) - \Phi(x)| < \frac{\varepsilon}{3\sqrt{|B_{2a}|}} \quad (x \in \mathbb{R}^n) \ \text{if} \ |z| < \delta.$$

Therefore,

$$\|\Phi(\cdot + z) - \Phi(\cdot)\|^2_{L_2(\mathbb{R}^n)} = \int_{B_{2a}} |\Phi(x+z) - \Phi(x)|^2\, dx < \frac{\varepsilon^2}{9|B_{2a}|}|B_{2a}| = \varepsilon^2/9.$$

The following inequality completes the proof:

$$\|f(\cdot + z) - f(\cdot)\|_{L_2(Q)} \leqslant \|F(\cdot + z) - F(\cdot)\|_{L_2(\mathbb{R}^n)} \leqslant \|F(\cdot + z) - \Phi(\cdot + z)\|_{L_2(\mathbb{R}^n)} + \\ + \|\Phi(\cdot + z) - \Phi(\cdot)\|_{L_2(\mathbb{R}^n)} + \|F - \Phi\|_{L_2(\mathbb{R}^n)} < \varepsilon/3 + \varepsilon/3 + \varepsilon/3 = \varepsilon.$$

□

Averaging

Fixing the mollifier $\omega_h(x)$, we assign to each function $f \in L_2(Q)$ the family of $C_0^\infty(\mathbb{R}^n)$-functions f_h by the formula

$$f_h(x) = \int_Q f(y)\omega_h(x-y)\,dy \quad (x \in \mathbb{R}^n;\ h > 0). \tag{4.2}$$

The set $\{f_h\}_{h>0}$ is called the *averaging* of the function f. The functions f_h are called *average functions*. One has $\operatorname{supp} f_h \subset \overline{Q^h}$, where $Q^h = \{x \in \mathbb{R}^n : \rho(x, Q) < h\}$.

Theorem 4.3. *The convergence* $f_h \overset{L_2(Q)}{\longrightarrow} f$, $h \to 0$, *takes place for any function* $f \in L_2(Q)$.

Proof. As usual, we assume that a function $f \in L_2(Q)$ is equal to zero outside Q. Using the properties of the mollifier (see Section 3, and the Cauchy–Schwarz inequality), we obtain the following estimate:

$$\begin{aligned} |f(x) - f_h(x)|^2 &= \left| f(x) \int_{\mathbb{R}^n} \omega_h(x-y)\,dy - \int_Q f(y)\omega_h(x-y)\,dy \right|^2 = \\ &= \left| \int_{|y-x|<h} (f(x) - f(y))\omega_h(x-y)\,dy \right|^2 \leqslant \\ &\leqslant \int_{|z|<h} \omega_h^2(z)\,dz \int_{|z|<h} |f(x+z) - f(x)|^2\,dz \leqslant \\ &\leqslant \frac{c_n}{eh^n} \int_{|z|<h} \omega_h(z)\,dz \int_{|z|<h} |f(x+z) - f(x)|^2\,dz = \\ &= \frac{c_n}{eh^n} \int_{|z|<h} |f(x+z) - f(x)|^2\,dz. \end{aligned}$$

Integrating the obtained inequality over Q and applying Fubini's theorem, we have

$$\|f - f_h\|^2_{L_2(Q)} \leqslant \frac{c_n}{eh^n} \int_{|z|<h} dz \int_Q |f(x+z) - f(x)|^2\,dx. \tag{4.3}$$

Let us fix $\varepsilon > 0$, then the existence of a number $\delta > 0$ such that the internal integral on the right-hand side of inequality (4.3) is less than ε for $h < \delta$ is guaranteed by Theorem 4.2. Finally,

$$\|f - f_h\|^2_{L_2(Q)} \leqslant \frac{\varepsilon c_n}{e h^n} \int\limits_{|z|<h} dz = \frac{c_n \sigma_n \varepsilon}{en} \quad \text{for } h < \delta. \qquad \square$$

Theorem 4.4. *The set $C_0^\infty(Q)$ is dense in $L_2(Q)$.*

Proof. Let us fix $f \in L_2(Q)$ and $\varepsilon > 0$. Since Q is bounded and $\partial Q \in C^0$, the absolute continuity of the Lebesgue integral allows us to choose $\delta > 0$ to ensure the inequality

$$\int\limits_{Q \setminus Q_\delta} |f(x)|^2 \, dx < \varepsilon^2/4$$

(recall that $Q_\delta = \{x \in Q : \rho(x, \partial Q) > \delta\}$). Let us introduce the truncated function

$$F(x) = \begin{cases} f(x), & x \in Q_\delta, \\ 0, & x \in Q \setminus Q_\delta, \end{cases}$$

and the corresponding averaging $F_h \in C_0^\infty(Q)$ for $h < \delta$. Reducing h appropriately, we have $\|F - F_h\|_{L_2(Q)} < \varepsilon/2$ by Theorem 4.3. Combining the inequalities, we get

$$\begin{aligned} \|f - F_h\|_{L_2(Q)} &\leqslant \|f - F\|_{L_2(Q)} + \|F - F_h\|_{L_2(Q)} = \\ &= \|f\|_{L_2(Q \setminus Q_\delta)} + \|F - F_h\|_{L_2(Q)} < \varepsilon/2 + \varepsilon/2 = \varepsilon. \end{aligned} \qquad \square$$

Theorem 4.5. *The space $L_2(Q)$ is separable.*

Proof. By Theorem 4.1, any function from $L_2(Q)$ can be approximated by a continuous function in $\overline{Q}$. On the other hand, since Q is a bounded domain, the Weierstrass theorem states that any function continuous in $\overline{Q}$ is the uniform limit of a sequence of polynomials in $x_1, \dots, x_n$ with rational coefficients. Note that uniform convergence implies convergence in the mean square, thus we conclude that the (countable) set of all polynomials with rational coefficients is dense everywhere in $L_2(Q)$. □

5 Weak derivatives

The concept of a weak derivative is closely related to the well-known formula of integration by parts. In its most abstract form, this concept is present in the theory of distributions (see, e. g., [15]). In this book, we restrict ourselves to the case where the functions themselves and their weak derivatives belong to $L_2(Q)$.

Definition 5.1. Let $L_{2,\mathrm{loc}}(Q)$ denote the set of all functions $u(x)$ measurable in Q such that $u \in L_2(\Omega)$ for any domain $\Omega \subset \mathbb{R}^n$, $\overline{\Omega} \subset Q$. Spaces $L_{p,\mathrm{loc}}(Q)$ are defined similarly for arbitrary $p \geqslant 1$.

Definition 5.2. Let α be a multiindex. The function $u_\alpha \in L_{2,\mathrm{loc}}(Q)$ is called the *weak derivative of order* α of a function $u \in L_{2,\mathrm{loc}}(Q)$ in the domain Q if the integral identity

$$\int_Q u(x)D^\alpha\varphi(x)\,dx = (-1)^{|\alpha|}\int_Q u_\alpha(x)\varphi(x)\,dx \tag{5.1}$$

is satisfied for any function $\varphi \in C_0^\infty(Q)$.

First of all, let us present some elementary consequences of relation (5.1).

1) The weak derivative is unique if it exists. Indeed, if $u_{\alpha,1}$ and $u_{\alpha,2}$ are weak derivatives of the function u in the domain Q, then from equation (5.1) it follows that

$$\int_Q (u_{\alpha,1}(x) - u_{\alpha,2}(x))\varphi(x)\,dx = 0 \quad (\varphi \in C_0^\infty(\Omega))$$

for any strictly internal subdomain Ω, $\overline{\Omega} \subset Q$. Hence, $u_{\alpha,1}(x) = u_{\alpha,2}(x)$ a. e. in Ω and, due to the arbitrariness of Ω, a. e. in Q.

2) The mixed derivative of a smooth function does not depend on the order of differentiation with respect to the variables $x_1, \dots, x_n$, so does the weak derivative u_α.

3) If u_α is the weak derivative of a function u in Q and $Q' \subset Q$ is a subdomain, then the restriction $u_\alpha|_{Q'}$ is the weak derivative of $u|_{Q'}$ in Q'.

4) If u is a compactly supported function from $L_2(Q)$, then its weak derivative u_α is compactly supported in Q as well. Moreover, $\operatorname{supp} u_\alpha \subset \operatorname{supp} u$.

5) Integrating by parts in the integral

$$\int_Q u(x)D^\alpha\varphi(x)\,dx$$

when $u \in C^k(Q)$, we see that $u_\alpha = D^\alpha u$ for $|\alpha| \leqslant k$, i. e., the weak derivative of a continuously-differentiable function coincides with the corresponding classical derivative. In what follows, the usual notation $D^\alpha u$ is used for weak derivatives.

Here, it is worth emphasizing the fundamental differences between the definition of the weak derivative and the classical definition of derivative.

First, the weak derivative is defined at once as a function in the domain. Its value at a given point has no meaning. It is a (locally) integrable function that can be changed as desired on a set of measure zero. Second, classical higher-order derivatives are determined inductively through lower-order derivatives. Thus, the second derivative $\partial^2 u/\partial x_1\partial x_2$ is, by definition, the derivative with respect to x_2 of the first derivative $\partial u/\partial x_1$. In the case of weak derivatives, the derivative corresponding to the given multiindex α is determined immediately. In this case, there is no need to worry about lower-order derivatives. Moreover, Example 5.2 shows that the weak derivative $\partial^2 u/\partial x_1\partial x_2$ can exist without the existence of first-order weak derivatives $\partial u/\partial x_1$ and $\partial u/\partial x_2$.

Example 5.1. Consider the function $u(x_1, x_2) = |x_1|$. Let us find the first-order weak derivatives u_{x_1}, u_{x_2} in $Q = \{x \in \mathbb{R}^2 : x_1^2 + x_2^2 < 1\}$.

Set $Q^+ = \{x \in Q : x_1 > 0\}$, $Q^- = \{x \in Q : x_1 < 0\}$, and let $\varphi \in C_0^\infty(Q)$. Then

$$\iint_Q u\varphi_{x_1}\,dx = \iint_{Q^+} x_1\varphi_{x_1}\,dx - \iint_{Q^-} x_1\varphi_{x_1}\,dx.$$

In the integrals over Q^+ and Q^- use integration by parts. Denoting by $\nu = (\nu_1, \nu_2)$ the unit vector of the outer normal to ∂Q^+, write

$$\iint_{Q^+} x_1\varphi_{x_1}\,dx = \int_{\partial Q^+} x_1\varphi\nu_1\,ds - \iint_{Q^+} \varphi\,dx = -\iint_{Q^+} \varphi\,dx,$$

taking into account the fact that $(x_1\varphi\nu_1)|_{\partial Q^+} = 0$. Similarly,

$$\iint_{Q^-} x_1\varphi_{x_1}\,dx = -\iint_{Q^-} \varphi\,dx.$$

As a result, we obtain

$$\iint_Q u\varphi_{x_1}\,dx = -\iint_Q \operatorname{sign} x_1\varphi\,dx,$$

i. e., the function $u_{x_1} = \operatorname{sign} x_1$ is the weak derivative of the function $u = |x_1|$. The same reasoning shows that u has the weak derivative u_{x_2} equal to zero.

Example 5.2. Let us show that the function $v(x_1, x_2) = \operatorname{sign} x_1$ has no weak derivative v_{x_1} from $L_2(Q)$, where Q is the same as in the previous example. Suppose that $w = v_{x_1} \in L_2(Q)$ is the weak derivative of the function v in Q. If the test function $\varphi \in C_0^\infty(Q)$ in the definition of w has a compact support in Q^+, then we obtain

$$0 = \iint_Q v\varphi_{x_1}\,dx = \iint_{Q^+} \varphi_{x_1}\,dx = -\iint_{Q^+} w\varphi\,dx,$$

whence it follows that $w|_{Q^+} = 0$. Similarly, if $\operatorname{supp}\varphi \subset Q^-$, then

$$0 = \iint_Q v\varphi_{x_1}\,dx = -\iint_{Q^-} \varphi_{x_1}\,dx = -\iint_{Q^-} w\varphi\,dx,$$

i. e., $w|_{Q^-} = 0$.

Therefore, $w = 0$ a. e. in Q and the integral on the right-hand side of the equality

$$\iint_Q v\varphi_{x_1}\,dx = -\iint_Q w\varphi\,dx$$

vanishes for any function $\varphi \in C_0^\infty(Q)$. However,

$$\iint_Q v\varphi_{x_1}\,dx = \iint_{Q^+} \varphi_{x_1}\,dx - \iint_{Q^-} \varphi_{x_1}\,dx = -2 \int_{Q\cap\{x_1=0\}} \varphi\,ds,$$

which leads to a contradiction.

Interestingly, the second-order mixed derivative $v_{x_1x_2}$ exists and is equal to zero. One could easily check that

$$\iint_Q v\varphi_{x_1x_2}\,dx = -2\int_{-1}^{1} \varphi_{x_2}(0,x_2)\,dx_2 = -2(\varphi(0,1) - \varphi(0,-1)) = 0$$

for $\varphi \in C_0^\infty(Q)$. This effect can be strengthened: the function $\operatorname{sign} x_1 + \operatorname{sign} x_2$ has the mixed second-order weak derivative equal to zero in the same domain Q, while both first-order derivatives do not exist.

Example 5.3. Let $Q = \{x \in \mathbb{R}^n : |x|^2 = x_1^2 + \cdots + x_n^2 < 1\}$. Let us find out for which values of a real parameter μ the function $u(x) = |x|^\mu$ has all first-order weak derivatives $u_{x_i} \in L_2(Q)$ $(i = 1,\ldots,n)$. First, we choose μ to ensure $u \in L_2(Q)$, i. e.,

$$\int_Q |x|^{2\mu}\,dx = \int_0^1 d\rho \int_{|x|=\rho} |x|^{2\mu}\,dS = \int_0^1 \rho^{2\mu}\,d\rho \int_{|x|=\rho} dS = \sigma_n \int_0^1 \rho^{2\mu+n-1}\,d\rho < \infty$$

if and only if $2\mu + n - 1 > -1$. The symbol σ_n denotes the surface area of the unit sphere in $\mathbb{R}^n$. We omit the trivial case $\mu = 0$.

Note that the function u is infinitely differentiable at all points except the origin, and

$$u_{x_i}(x) = \mu x_i |x|^{\mu-2} \quad (i = 1,\ldots,n;\ x \neq 0). \tag{5.2}$$

These functions obviously belong to $L_2(Q)$ if and only if $\mu > 1 - n/2$. It remains to check the integral identity. We use the following representation for the corresponding integral on the left-hand side of identity (5.1):

$$\int_Q u\varphi_{x_i}\,dx = \int_{|x|<\varepsilon} u\varphi_{x_i}\,dx + \int_{\varepsilon<|x|<1} u\varphi_{x_i}\,dx, \tag{5.3}$$

and the first term tends to zero as $\varepsilon \to 0$ due to the absolute continuity of the Lebesgue integral. Integrating by parts in the second term gives (note that $\nu = -x/\varepsilon$ on the sphere $|x| = \varepsilon$ and $\operatorname{supp}\varphi \subset Q$)

$$\int_{\varepsilon<|x|<1} u\varphi_{x_i}\,dx = -\frac{1}{\varepsilon}\int_{|x|=\varepsilon} x_i u\varphi\,dS - \int_{\varepsilon<|x|<1} u_{x_i}\varphi\,dx. \tag{5.4}$$

If $C = \max_{x\in\overline{Q}} |\varphi(x)|$, then

$$\left|\frac{1}{\varepsilon}\int\limits_{|x|=\varepsilon} x_i u\varphi\, dS\right| \leqslant C\varepsilon^{\mu}\int\limits_{|x|=\varepsilon} dS = C\sigma_n\varepsilon^{\mu+n-1} \to 0, \quad \varepsilon \to 0$$

since $\mu + n - 1 > 0$. Combining (5.3) and (5.4) and passing to the limit as $\varepsilon \to 0$, we obtain the equalities

$$\int\limits_Q u\varphi_{x_i}\, dx = -\int\limits_Q u_{x_i}\varphi\, dx$$

with functions u_{x_i} given by (5.2). Therefore, the function $u(x) = |x|^{\mu}$ has the first-order weak derivatives from $L_2(Q)$ if and only if $\mu > 1 - n/2$.

Averaging of weak derivatives

Theorem 5.1. *Let $D^{\alpha}f \in L_2(Q)$ be the weak derivative of a function $f \in L_2(Q)$ and f_h be determined by formula* (4.2). *As $x \in Q$, the equality*

$$D^{\alpha}f_h(x) = (D^{\alpha}f)_h(x) \tag{5.5}$$

holds for all $h < \rho(x, \partial Q)$. Moreover,

$$\|D^{\alpha}f_h - D^{\alpha}f\|_{L_2(\Omega)} \to 0, \quad h \to 0, \tag{5.6}$$

for any strictly internal subdomain Ω, $\overline{\Omega} \subset Q$.

If, in addition, f is compactly supported in Q and $h < \rho(\operatorname{supp} f, \partial Q)/2$, then equation (5.5) *holds for all points $x \in Q$ and convergence in* (5.6) *occurs for $\Omega = Q$.*

Proof. Differentiating the integral $\int_Q f(y)\omega_h(x-y)dy$ with respect to x under the integral sign, we obtain

$$D^{\alpha}f_h(x) = \int\limits_Q f(y)D_x^{\alpha}\omega_h(x-y)\, dy = (-1)^{|\alpha|}\int\limits_Q f(y)D_y^{\alpha}\omega_h(x-y)\, dy.$$

Note that $\omega_h(x - \cdot) \in C_0^{\infty}(Q)$ if $h < \rho(x, \partial Q)$. Therefore, according to the definition of weak derivative, the expression on the right-hand side is equal to

$$(-1)^{2|\alpha|}\int\limits_Q D^{\alpha}f(y)\omega_h(x-y)\, dy = (D^{\alpha}f)_h(x).$$

Let $\overline{\Omega} \subset Q$, thus $\rho(\overline{\Omega}, \partial Q) > 0$. If $h < \rho(\overline{\Omega}, \partial Q)$, then we already have $D^{\alpha}f_h(x) = (D^{\alpha}f)_h(x)$ for all $x \in \overline{\Omega}$. Together with Theorem 4.3 applied to the weak derivative $D^{\alpha}f$, this gives

$$\|D^\alpha f_h - D^\alpha f\|_{L_2(\Omega)} = \|(D^\alpha f)_h - D^\alpha f\|_{L_2(\Omega)} \to 0, \quad h \to 0.$$

In order to complete the proof in the case of $\operatorname{supp} f \subset Q$, it is enough to note that all functions $D^\alpha f, D^\alpha f_h$ and $(D^\alpha f)_h$ are supported in Q_δ for $h < \delta$, where $\delta = \rho(\operatorname{supp} f, \partial Q)/2$. The statement of the theorem follows from what was proved above. □

Theorem 5.2. *Let a function $f \in L_2(Q)$ be such that all of its first-order weak derivatives are equal to zero in Q. Then f is constant in Q.*

Proof. Take any strictly internal subdomain Ω. It follows from Theorem 5.1 that $(f_h)_{x_i}(x) = (f_{x_i})_h(x) = 0$ in Ω as long as h is small enough. Since $f_h \in C^\infty(\overline{Q})$, the last relation guarantees its constancy in Ω. However, the corresponding constant may depend on h. Let us denote it by c_h. Theorem 4.3 states that the family $\{f_h\}$ converges to the function f in $L_2(\Omega)$ as $h \to 0$, and thus

$$\|f_{h_1} - f_{h_2}\|_{L_2(\Omega)} = |\Omega|^{1/2}|c_{h_1} - c_{h_2}| \to 0, \quad h_1, h_2 \to 0.$$

Therefore, there exists the limit $c = \lim_{h\to 0} c_h \in \mathbb{C}$, and the equality $f|_\Omega = c$ holds. But the subdomain Ω, $\overline{\Omega} \subset Q$, is arbitrary, thus $f = c$ in Q. □

Weak derivative in a union of domains

First of all, we give a useful criterion for the existence of weak derivative.

Theorem 5.3. *Let $f \in L_2(Q)$, f_h be the averaging of f, and α a multiindex. If the set $\{D^\alpha f_h\}$ is bounded in $L_2(\Omega)$ for any strictly interior subdomain Ω, $\overline{\Omega} \subset Q$, and all sufficiently small h, $h < h(\Omega)$, then the weak derivative $D^\alpha f \in L_{2,\mathrm{loc}}(Q)$ exists in Q.*

Proof. Let us take an arbitrary sequence of subdomains Q_i, $i = 1, 2, \dots$ such that $\overline{Q}_i \subset Q_{i+1} \subset Q$ and $Q = \cup Q_i$.

Being bounded in $L_2(Q_1)$, the set $\{D^\alpha f_h\}$ is weakly compact for small $h > 0$, i. e., there exists a sequence $h_{1,k} \to 0$ such that the sequence $D^\alpha f_{h_{1,k}}$ weakly converges in $L_2(Q_1)$. But this sequence is also bounded in $L_2(Q_2)$. Therefore, we can extract a subsequence $D^\alpha f_{h_{2,k}}$ converging weakly in $L_2(Q_2)$, etc. As a result of this procedure, we obtain a diagonal sequence $D^\alpha f_{h_{k,k}}$ that weakly converges in $L_2(Q_i)$ for all $i = 1, 2, \dots$. We denote the corresponding weak limit by $\omega \in L_{2,\mathrm{loc}}(Q)$. Obviously, $D^\alpha f_{h_{k,k}}$ converges to ω weakly in $L_2(\Omega)$ for any strictly interior subdomain Ω.

Now let a function $g \in C_0^\infty(Q)$ and a subdomain Ω be such that $\overline{\Omega} \subset Q$ and g vanish outside Ω. The equality

$$\int_Q D^\alpha f_{h_{k,k}} \overline{g}\, dx = (-1)^{|\alpha|} \int_Q f_{h_{k,k}} D^\alpha \overline{g}\, dx \tag{5.7}$$

is then valid for all $k = 1, 2, \ldots$. Note that in fact the integration in (5.7) occurs over Ω. Since $D^\alpha f_{h_{k,k}}$ converges to ω weakly in $L_2(\Omega)$ while the sequence $f_{h_{k,k}}$ itself converges to f with respect to the $L_2(Q)$-norm, we can pass to the limit as $k \to \infty$ in (5.7) to obtain the relation

$$\int_Q \omega \overline{g}\, dx = (-1)^{|\alpha|} \int_Q f D^\alpha \overline{g}\, dx,$$

which means that $\omega \in L_{2,\mathrm{loc}}(Q)$ is the weak derivative $D^\alpha f$ of the function f in Q. □

Note that the converse statement follows from Theorem 5.1.

Let us now prove the main result of this section.

Theorem 5.4. *Let domains Q_1 and Q_2 be such that their union $Q = Q_1 \cup Q_2$ is also a domain. Given a function $f \in L_2(Q)$, we denote $f_1 = f|_{Q_1}$ and $f_2 = f|_{Q_2}$. If the weak derivatives $D^\alpha f_i \in L_2(Q_i)$ $(i = 1, 2)$ exist, then the function f has the weak derivative $D^\alpha f \in L_2(Q)$.*

Proof. For each point $x \in Q$, choose its neighborhood U (an open ball centered at x) as follows: if $x \in Q_1$, then require $\overline{U} \subset Q_1$, and if $x \in Q_2 \setminus Q_1$, then $\overline{U} \subset Q_2$.

Choose an arbitrary domain Ω, $\overline{\Omega} \subset Q$. We have a cover of the compact set $\overline{\Omega}$ by the open balls U as x runs through Q. Select a finite subcover $U_1, \ldots, U_N$. Denote by Ω_1 the union of all balls U_j whose centers belong to Q_1. By construction, $\overline{\Omega}_1 \subset Q_1$. The centers of the remaining balls U_j are in $Q_2 \setminus Q_1$. For their union Ω_2, we have $\overline{\Omega}_2 \subset Q_2$. Moreover, $\Omega \subset \Omega_1 \cup \Omega_2$.

Since functions f_i have weak derivatives $D^\alpha f_i \in L_2(Q_i)$, the set $\{D^\alpha f_h\}$ is bounded in $L_2(\Omega_i)$ for all small h $(i = 1, 2)$, where f_h are the averaged functions for f in Q (note that it does not matter whether we average over Q_i or over the entire domain Q, once we are only interested in the points of Ω_i and h is quite small). Due to the obvious inequality,

$$\|D^\alpha f_h\|^2_{L_2(\Omega)} \leqslant \|D^\alpha f_h\|^2_{L_2(\Omega_1)} + \|D^\alpha f_h\|^2_{L_2(\Omega_2)},$$

the set $\{D^\alpha f_h\}$ is bounded in $L_2(\Omega)$. Since Ω is arbitrary, the existence of weak derivative $D^\alpha f$ follows from Theorem 5.3. In each of the domains Q_i $(i = 1, 2)$, $D^\alpha f$ coincides with $D^\alpha f_i$ and, therefore, belongs to $L_2(Q)$. □

Approximation of weak derivatives by finite differences

The results of this section play an essential role in the study of smoothness of weak solutions to elliptic boundary-value problems in Chapter 3.

For an arbitrary function $f \in L_2(Q)$ extended by zero to $\mathbb{R}^n \setminus Q$ and any real number $h \neq 0$, introduce the ratio

$$\delta_h^j f(x) = \frac{f(x_1, \ldots, x_{j-1}, x_j + h, x_{j+1}, \ldots, x_n) - f(x)}{h}.$$

Obviously, $\delta_h^j f \in L_2(Q)$. Moreover, $\operatorname{supp} \delta_h^j f \subset Q$ when $\operatorname{supp} f \subset Q$ and h is small. If g is another function from $L_2(Q)$ extended by zero to $\mathbb{R}^n \setminus Q$, then the finite-difference analogue

$$(\delta_h^j f, g)_{L_2(Q)} = -(f, \delta_{-h}^j g)_{L_2(Q)} \tag{5.8}$$

of integration by parts formula holds.

Theorem 5.5. *Let Q be a bounded domain in $\mathbb{R}^n$, f be a compactly supported function in $L_2(Q)$, and $j \in \{1, \dots, n\}$.*

(a) *If there exists the weak derivative $f_{x_j} \in L_2(Q)$, then*

$$\|\delta_h^j f\|_{L_2(Q)} \leqslant \|f_{x_j}\|_{L_2(Q)} \tag{5.9}$$

for all sufficiently small numbers $h \neq 0$, and

$$\|\delta_h^j f - f_{x_j}\|_{L_2(Q)} \to 0 \quad \text{as } h \to 0. \tag{5.10}$$

(b) *If there exists a constant $C > 0$ such that the inequality $\|\delta_h^j f\|_{L_2(Q)} \leqslant C$ holds for all sufficiently small $h \neq 0$, then the function f has the weak derivative $f_{x_j} \in L_2(Q)$, and the estimate $\|f_{x_j}\|_{L_2(Q)} \leqslant C$ and relation* (5.10) *hold.*

Proof. First, let $f \in C_0^1(Q)$. We can assume without loss of generality that $j = n$, $x = (x', x_n)$ and $h > 0$. Expressing the difference $\delta_h^n f$ by the Newton–Leibniz formula,

$$\delta_h^n f(x) = \frac{1}{h} \int_{x_n}^{x_n+h} f_{x_n}(x', \xi)\, d\xi,$$

and applying the Cauchy–Schwarz inequality, we get

$$|\delta_h^n f(x)|^2 \leqslant \frac{1}{h^2} \left(\int_{x_n}^{x_n+h} |f_{x_n}(x', \xi)|\, d\xi \right)^2 \leqslant \frac{1}{h} \int_{x_n}^{x_n+h} |f_{x_n}(x', \xi)|^2\, d\xi.$$

Integrating this inequality over x_n and using Fubini's theorem, we obtain

$$\begin{aligned} \int_{-\infty}^{+\infty} |\delta_h^n f(x)|^2\, dx_n &\leqslant \frac{1}{h} \int_{-\infty}^{+\infty} dx_n \int_{x_n}^{x_n+h} |f_{x_n}(x', \xi)|^2\, d\xi = \\ &= \frac{1}{h} \int_{-\infty}^{+\infty} d\xi \int_{\xi-h}^{\xi} |f_{x_n}(x', \xi)|^2\, dx_n = \int_{-\infty}^{+\infty} |f_{x_n}(x', x_n)|^2\, dx_n. \end{aligned}$$

It remains to integrate the last inequality over $x' \in \mathbb{R}^{n-1}$ to obtain estimate (5.9).

Further,

$$\delta_h^n f(x) - f_{x_n}(x) = \frac{1}{h}\int\limits_{x_n}^{x_n+h} f_{x_n}(x',\xi)\,d\xi - f_{x_n}(x) = \\ = \frac{1}{h}\int\limits_{x_n}^{x_n+h} (f_{x_n}(x',\xi) - f_{x_n}(x',x_n))\,d\xi.$$

Using the Cauchy–Schwarz inequality and then Fubini's theorem, we deduce that

$$\int\limits_{-\infty}^{+\infty} |\delta_h^n f(x) - f_{x_n}(x)|^2\,dx_n \leqslant \frac{1}{h}\int\limits_{-\infty}^{+\infty} dx_n \int\limits_{x_n}^{x_n+h} |f_{x_n}(x',\xi) - f_{x_n}(x',x_n)|^2\,d\xi = \\ = \frac{1}{h}\int\limits_{0}^{h} d\eta \int\limits_{-\infty}^{+\infty} |f_{x_n}(x',x_n+\eta) - f_{x_n}(x',x_n)|^2\,dx_n$$

and, finally,

$$\|\delta_h^n f - f_{x_n}\|^2_{L_2(Q)} \leqslant \frac{1}{h}\int\limits_{0}^{h} d\eta \int\limits_{Q} |f_{x_n}(x',x_n+\eta) - f_{x_n}(x',x_n)|^2\,dx. \tag{5.11}$$

Estimates (5.9) and (5.11) are obtained for all compactly supported continuously differentiable functions in Q.

Now, for an arbitrary compactly supported function f from $L_2(Q)$ possessing the weak derivative $f_{x_n} \in L_2(Q)$, consider the averaged functions $f_\rho(x) = \int_Q f(y)\omega_\rho(x-y)\,dy$ lying in $C_0^1(Q)$ if $\rho > 0$ is small enough, for which we have $f_\rho \xrightarrow{L_2(Q)} f$ and $(f_\rho)_{x_n} \xrightarrow{L_2(Q)} f_{x_n}$, $\rho \to 0$, by Theorem 5.1. Substituting f_ρ for f in (5.9) and (5.11) and passing to the limit as $\rho \to 0$, we establish the same inequalities for f.

For any number $\varepsilon > 0$, Theorem 4.2 guarantees the inequality

$$\int\limits_{Q} |f_{x_n}(x',x_n+\eta) - f_{x_n}(x',x_n)|^2\,dx < \varepsilon^2 \quad (0 \leqslant \eta \leqslant h)$$

for all sufficiently small h. Consequently, the integral on the right-hand side of equation (5.11) turns out to be less than ε^2. This proves relation (5.10) and statement (a) of the theorem.

Now let us prove statement (b). Being bounded in the Hilbert space $L_2(Q)$, the set $\{\delta_h^j f\}$ is weakly compact in $L_2(Q)$ for small h. Therefore, we take a weakly convergent subsequence $\delta_{h_p}^j f$, $p = 1, 2, \ldots$, for which $h_p \to 0$ as $p \to \infty$. Let us denote this weak limit by ω. Obviously, $\|\omega\|_{L_2(Q)} \leqslant C$. Fix an arbitrary function $g \in C_0^\infty(Q)$ in the equality

$$(\delta^j_{h_p} f, g)_{L_2(Q)} = -(f, \delta^j_{-h_p} g)_{L_2(Q)}$$

(see formula (5.8)). Lebesgue's dominated convergence theorem is applicable to the integral on the right-hand side, according to which

$$(f, \delta^j_{-h_p} g)_{L_2(Q)} \to (f, g_{x_j})_{L_2(Q)} \quad \text{as } p \to \infty.$$

On the other hand, the integral on the left-hand side converges to $(\omega, g)_{L_2(Q)}$, thus we have $(\omega, g)_{L_2(Q)} = -(f, g_{x_j})_{L_2(Q)}$ for $g \in C_0^\infty(Q)$. We have shown that the function $\omega \in L_2(Q)$ is the weak derivative of f_{x_j}. This concludes the proof. □

The following result is a significant addition to Theorem 5.5.

Let Q be a simply connected bounded domain symmetric with respect to the plane $x_n = 0$ and let a positive number δ be such that the set Q_δ is a domain. Let us denote

$$Q^+ = Q \cap \{x_n > 0\}, \quad Q^- = Q \cap \{x_n < 0\}, \quad (Q_\delta)^+ = Q_\delta \cap \{x_n > 0\}.$$

Theorem 5.6. *Let $f \in L_2(Q^+)$ and $f(x) = 0$ in $Q^+ \setminus (Q_\delta)^+$.*

(a) *If there exists the weak derivative $f_{x_j} \in L_2(Q^+)$, $j < n$, then*

$$\|\delta^j_h f\|_{L_2(Q^+)} \leqslant \|f_{x_j}\|_{L_2(Q^+)}$$

for all sufficiently small $h \neq 0$, and

$$\|\delta^j_h f - f_{x_j}\|_{L_2(Q^+)} \to 0 \quad \text{as } h \to 0. \tag{5.12}$$

(b) *If there exists a constant $C > 0$ such that the inequality $\|\delta^j_h f\|_{L_2(Q^+)} \leqslant C$ with $j < n$ holds for all sufficiently small $h \neq 0$, then the function f has the weak derivative $f_{x_j} \in L_2(Q^+)$ in Q^+, and the estimate $\|f_{x_j}\|_{L_2(Q^+)} \leqslant C$ and relation* (5.12) *hold.*

Proof. For each function f in Q^+ satisfying the conditions of the theorem we define its even extension F in Q by setting $F(x) = f(x', -x_n)$ in Q^-. It is clear that $F \in L_2(Q)$ and $F(x) = 0$ for $x \notin Q_\delta$, i. e., the function F is compactly supported in Q. In addition, $\|\delta^j_h F\|^2_{L_2(Q)} = 2\|\delta^j_h f\|^2_{L_2(Q^+)}$ when $0 < |h| < \delta$.

Under the conditions of statement (a) of the theorem, we verify the existence of weak derivative F_{x_j} from $L_2(Q)$. Choose any function $\zeta_r \in C^\infty(\mathbb{R})$, $r > 0$, such that $\zeta_r(-t) = \zeta_r(t)$, $0 \leqslant \zeta_r(t) \leqslant 1$, $\zeta_r(t) = 0$ for $|t| \leqslant r/2$, and $\zeta_r(t) = 1$ for $|t| \geqslant r$. For $g \in C_0^\infty(Q)$, we obviously have

$$\int_Q F(x) g_{x_j}(x) \zeta_r(x_n)\, dx =$$

$$= \int_{Q^+} f(x) g_{x_j}(x) \zeta_r(x_n)\, dx + \int_{Q^-} f(x', -x_n) g_{x_j}(x) \zeta_r(x_n)\, dx =$$

$$= \int_{Q^+} f(x)[\zeta_r(x_n)(g(x',x_n)+g(x',-x_n))]_{x_j}\,dx.$$

Since the function $\zeta_r(x_n)(g(x',x_n)+g(x',-x_n))$ belongs to $C_0^\infty(Q^+)$, we can apply the definition of weak derivative:

$$\int_Q F(x)g_{x_j}(x)\zeta_r(x_n)\,dx = -\int_{Q^+} f_{x_j}(x)\zeta_r(x_n)(g(x',x_n)+g(x',-x_n))\,dx =$$
$$= -\int_{Q^+} f_{x_j}(x)\zeta_r(x_n)g(x)\,dx - \int_{Q^-} f_{x_j}(x',-x_n)\zeta_r(x_n)g(x)\,dx.$$

Passing to the limit as $r \to 0$ in the last relation and using Lebesgue's dominated convergence theorem, we obtain the equality

$$\int_Q Fg_{x_j}\,dx = -\int_{Q^+} f_{x_j}g\,dx - \int_{Q^-} f_{x_j}(x',-x_n)g\,dx.$$

This means that the function equal to f_{x_j} in Q^+ and $f_{x_j}(x',-x_n)$ in Q^- is the weak derivative of F_{x_j}, and $\|F_{x_j}\|^2_{L_2(Q)} = 2\|f_{x_j}\|^2_{L_2(Q^+)}$.

Therefore, the function F satisfies the conditions of statement (a) of Theorem 5.5. It gives

$$\|\delta_h^j f\|^2_{L_2(Q^+)} = \|\delta_h^j F\|^2_{L_2(Q)}/2 \leqslant \|F_{x_j}\|^2_{L_2(Q)}/2 = \|f_{x_j}\|^2_{L_2(Q^+)}.$$

Moreover,

$$\|\delta_h^j f - f_{x_j}\|^2_{L_2(Q^+)} = \|\delta_h^j F - F_{x_j}\|^2_{L_2(Q)}/2 \to 0, \quad h \to 0.$$

Statement (a) is now proven.

We have $\|\delta_h^j F\|^2_{L_2(Q)} \leqslant 2C^2$ under the conditions of statement (b), whence it follows by Theorem 5.5 that the function F has the weak derivative in Q, and $\|F_{x_j}\|^2_{L_2(Q)} \leqslant 2C^2$. But then f also has a weak derivative in Q^+, and the estimate $\|f_{x_j}\|^2_{L_2(Q^+)} \leqslant C^2$ holds. This completes the proof. □

6 Sobolev spaces $H^k(Q)$

In this section, k is a fixed natural number.

Definition 6.1. The Sobolev space $H^k(Q)$ is the linear space of all complex-valued functions from $L_2(Q)$ having all weak derivatives up to order k inclusively from $L_2(Q)$,

$$H^k(Q) = \{u \in L_2(Q) : D^\alpha u \in L_2(Q),\ |\alpha| \leqslant k\}.$$

In the space $H^k(Q)$, the inner product is introduced by the formula

$$(u, v)_{H^k(Q)} = \sum_{|\alpha|\leqslant k} (D^\alpha u, D^\alpha v)_{L_2(Q)}.$$

The corresponding norm is defined as

$$\|u\|_{H^k(Q)} = \Big(\sum_{|\alpha|\leqslant k} \|D^\alpha u\|^2_{L_2(Q)} \Big)^{1/2}. \tag{6.1}$$

Obviously,

$$(u, v)_{H^1(Q)} = (u, v)_{L_2(Q)} + \sum_{i=1}^{n} (u_{x_i}, v_{x_i})_{L_2(Q)} = (u, v)_{L_2(Q)} + (\nabla u, \nabla v)_{L_2^n(Q)}$$

and

$$\|u\|_{H^1(Q)} = \big(\|u\|^2_{L_2(Q)} + \||\nabla u|\|^2_{L_2(Q)}\big)^{1/2},$$

where $L_2^n(Q) = L_2(Q) \times \cdots \times L_2(Q)$, $\nabla u = (u_{x_1}, \ldots, u_{x_n})$.

Convergence in $H^k(Q)$ can be described as the mean-square convergence for functions themselves and for all their weak derivatives up to order k. Obviously, $C^k(\overline{Q}) \subset H^k(Q)$. Moreover, the space $C^k(\overline{Q})$ is continuously embedded in the Sobolev space $H^k(Q)$ since the uniform convergence implies convergence in the mean square.

Theorem 6.1. *The space $H^k(Q)$ is complete with respect to norm* (6.1), *i. e., is a Hilbert space.*

Proof. Let $\{f_m\}_{m=1}^\infty$ be a Cauchy sequence in $H^k(Q)$,

$$\|f_m - f_p\|_{H^k(Q)} \to 0, \quad m, p \to \infty.$$

It means that

$$\|D^\alpha f_m - D^\alpha f_p\|_{L_2(Q)} \to 0, \quad m, p \to \infty$$

for all multiindices α, $|\alpha| \leqslant k$, i. e., each sequence $\{D^\alpha f_m\}_{m=1}^\infty$ is a Cauchy sequence in $L_2(Q)$. The completeness of the space $L_2(Q)$ implies the existence of limit functions $f_0 \in L_2(Q)$ and $f_\alpha \in L_2(Q)$, $|\alpha| \leqslant k$, such that

$$\|f_0 - f_m\|_{L_2(Q)} \to 0, \quad \|f_\alpha - D^\alpha f_m\|_{L_2(Q)} \to 0, \quad m \to \infty. \tag{6.2}$$

Let us show that each function f_α is the weak αth derivative of the function f_0. For $\varphi \in C_0^\infty(Q)$, we have

$$(f_m, D^\alpha \varphi)_{L_2(Q)} = (-1)^{|\alpha|}(D^\alpha f_m, \varphi)_{L_2(Q)}$$

(replacing φ with the complex-conjugate function $\overline{\varphi}$ in integral identity (5.1) does not affect the definition of weak derivative). Since the inner product is continuous with respect to its arguments, we can pass to the limit as $m \to \infty$ and obtain the equality

$$(f_0, D^\alpha \varphi)_{L_2(Q)} = (-1)^{|\alpha|}(f_\alpha, \varphi)_{L_2(Q)},$$

which means that the function f_α is the weak derivative of the function f_0, $f_\alpha = D^\alpha f_0$. Now it follows from (6.2) that

$$\|D^\alpha f_0 - D^\alpha f_m\|_{L_2(Q)} \to 0, \quad m \to \infty \quad (|\alpha| \leqslant k),$$

i. e., $\|f_0 - f_m\|_{H^k(Q)} \to 0, m \to \infty$. □

Some of the properties of the spaces $H^k(Q)$ are listed below. The reader is invited to reconstruct the proofs as an exercise.

1) If $f \in H^k(Q)$, then $f \in H^k(\Omega)$ for any subdomain $\Omega \subset Q$.

2) Let domains Q_1, Q_2 and Q be such that $Q = Q_1 \cup Q_2$, and a function f be defined in Q. If $f|_{Q_i} \in H^k(Q_i)$, $i = 1, 2$, then $f \in H^k(Q)$.

3) If $f \in H^k(Q)$ and $a \in C^k(\overline{Q})$, then $af \in H^k(Q)$, and for weak derivatives $D^\alpha(af)$ the Leibniz formula for differentiation of a product holds, e. g.,

$$(af)_{x_i} = a_{x_i} f + a f_{x_i} \quad (i = 1, \dots, n).$$

Moreover, multiplication by a function a turns out to be a bounded operator in $H^k(Q)$, i. e.,

$$\|af\|_{H^k(Q)} \leqslant c\|f\|_{H^k(Q)},$$

where the constant $c > 0$ does not depend on $f \in H^k(Q)$.

4) Let domains Q and Ω be such that $\overline{\Omega} \subset Q, f \in H^k(Q)$, and $\{f_h\}$ be the averaging of f; see (4.2). Then $\|f_h - f\|_{H^k(\Omega)} \to 0$ as $h \to 0$. If, in addition, $\operatorname{supp} f \subset Q$, then $\|f_h - f\|_{H^k(Q)} \to 0$ as $h \to 0$.

5) Suppose that $y = y(x)$ is a one-to-one mapping of a domain Q onto a domain Ω, its coordinate functions $y_i(x)$ belong to $C^k(\overline{Q})$, and the coordinate functions $x_i(y)$ of the inverse map belong to $C^k(\overline{\Omega})$. To each function $f(x)$ in Q we associate the function $F(y) = f(x(y))$ in Ω.

The function F belongs to $H^k(\Omega)$ if and only if the function f belongs to $H^k(Q)$, and for $D^\alpha F$ the usual chain rule holds. For example, the first-order weak derivatives are expressed by the formula

$$F_{y_i}(y) = \sum_{j=1}^{n} f_{x_j}(x(y)) \frac{\partial x_j}{\partial y_i} \quad (i = 1, \dots, n).$$

Moreover, there exist positive constants c_1 and c_2 such that the inequalities

$$c_1\|F\|_{H^k(\Omega)} \leqslant \|f\|_{H^k(Q)} \leqslant c_2\|F\|_{H^k(\Omega)}$$

hold for all $f \in H^k(Q)$.

Let us now move on to less obvious properties of the space $H^k(Q)$.

Extension theorem. Density of $C^\infty(\overline{Q})$ in $H^k(Q)$

Let Q be the cube $K_a = \{x \in \mathbb{R}^n : |x_i| < a\ (i = 1, \dots, n)\}$.

Lemma 6.1. *The set $C^\infty(\overline{K}_a)$ is dense in $H^k(K_a)$.*

Proof. Consider $\sigma > 1$. To any function $f \in H^k(K_a)$, we associate the function f_σ in the dilated cube $K_{\sigma a}$ by the formula $f_\sigma(x) = f(x/\sigma)$, $x \in K_{\sigma a}$. Obviously, $f_\sigma \in H^k(K_{\sigma a})$. Consider the averaging $\{(f_\sigma)_h\}$, consisting of functions that are smooth in $\mathbb{R}^n$. Due to the triangle inequality, one has

$$\|f - (f_\sigma)_h\|_{H^k(K_a)} \leqslant \|f - f_\sigma\|_{H^k(K_a)} + \|f_\sigma - (f_\sigma)_h\|_{H^k(K_a)}.$$

For any number $\sigma > 1$, the second term on the right-hand side of this inequality can be made arbitrarily small by choosing h close to zero (see property 4 of Sobolev spaces, taking into account that the cube K_a is strictly internal to $K_{\sigma a}$). We are going to estimate

$$\|f - f_\sigma\|^2_{H^k(K_a)} = \sum_{|\alpha|\leqslant k} \|D^\alpha f - D^\alpha f_\sigma\|^2_{L_2(K_a)}.$$

For every multiindex α, $|\alpha| \leqslant k$, and any number $\varepsilon > 0$, the existence of a function $\varphi_\alpha \in C(\overline{K}_a)$ such that $\|D^\alpha f - \varphi_\alpha\|_{L_2(K_a)} < \varepsilon$ is guaranteed by Theorem 4.1. Introducing $\varphi_{\alpha\sigma}(x) = \varphi_\alpha(x/\sigma)$, $x \in K_{\sigma a}$, and applying the triangle inequality, we obtain

$$\begin{aligned}&\|D^\alpha f - D^\alpha f_\sigma\|_{L_2(K_a)} \leqslant \\ &\quad \leqslant \|D^\alpha f - \varphi_\alpha\|_{L_2(K_a)} + \|\varphi_\alpha - \varphi_{\alpha\sigma}\|_{L_2(K_a)} + \|\varphi_{\alpha\sigma} - D^\alpha f_\sigma\|_{L_2(K_a)}.\end{aligned}$$

The uniform continuity of the function φ_α implies the uniform smallness of the difference $\varphi_\alpha - \varphi_{\alpha\sigma}$ in $\overline{K}_a$. Therefore, the second term of the right-hand side of the last inequality can be made arbitrarily small by choosing σ close to 1. Considering the third term, we see that

$$\begin{aligned}\|\varphi_{\alpha\sigma} - D^\alpha f_\sigma\|^2_{L_2(K_a)} &= \int_{K_a} |\varphi_{\alpha\sigma}(x) - D^\alpha f_\sigma(x)|^2\, dx \leqslant \\ &\leqslant \int_{K_{\sigma a}} |\varphi_\alpha(x/\sigma) - D^\alpha_x(f(x/\sigma))|^2\, dx = \sigma^n \int_{K_a} |\varphi_\alpha(y) - \sigma^{-|\alpha|} D^\alpha_y f(y)|^2 dy,\end{aligned}$$

and

$$\|\varphi_{\alpha\sigma} - D^\alpha f_\sigma\|_{L_2(K_a)} \leqslant \sigma^{n/2}\|\varphi_\alpha - \sigma^{-|\alpha|}D^\alpha f\|_{L_2(K_a)} \leqslant$$
$$\leqslant \sigma^{n/2}\|\varphi_\alpha - D^\alpha f\|_{L_2(K_a)} + \sigma^{n/2}(1-\sigma^{-|\alpha|})\|D^\alpha f\|_{L_2(K_a)} < 2\varepsilon$$

if σ is close to 1.

We have finally shown that any function $f \in H^k(K_a)$ can be approximated in the space $H^k(K_a)$ by functions $(f_\sigma)_h$. □

We use the notation $x = (x', x_n)$, $x' = (x_1, \dots, x_{n-1})$,

$$K_a^+ = K_a \cap \{x_n > 0\}, \quad K_a^- = K_a \cap \{x_n < 0\}.$$

Lemma 6.2. *Given an arbitrary function $f \in C^k(\overline{K_a^+})$, one can construct a function $F \in C^k(\overline{K_a})$ such that $F|_{K_a^+} = f$ and*

$$\|F\|_{C^k(\overline{K_a})} \leqslant M\|f\|_{C^k(\overline{K_a^+})},$$

where M is a positive constant independent of f. The statement remains true if we replace C^k with the space H^k.

Proof. Consider the system of linear algebraic equations

$$\sum_{i=1}^{k+1} A_i(-1/i)^s = 1 \quad (s = 0, \dots, k). \tag{6.3}$$

The determinant of this system is the Vandermonde determinant and, therefore, is nonzero. Thus, the set of numbers $A_1, \dots, A_{k+1}$ is uniquely determined by system (6.3). Taking $f \in C^k(\overline{K_a^+})$, we define the function $F(x)$ in K_a as follows:

$$F(x) = \begin{cases} f(x), & x \in \overline{K_a^+}, \\ \sum_{i=1}^{k+1} A_i f(x', -x_n/i), & x \in \overline{K_a^-}. \end{cases}$$

Obviously, $F|_{\overline{K_a^-}} \in C^k(\overline{K_a^-})$. Make sure that

$$\lim_{\substack{x \to (y',0),\\ x_n < 0}} D^\alpha F(x', x_n) = D^\alpha f(y', 0) \quad (|\alpha| \leqslant k).$$

Based on the form of F and the first equation ($s = 0$) of system (6.3), we see that

$$\lim_{\substack{x \to (y',0),\\ x_n < 0}} F(x', x_n) = \lim_{\substack{x \to (y',0),\\ x_n < 0}} \sum_{i=1}^{k+1} A_i f(x', -x_n/i) = \sum_{i=1}^{k+1} A_i f(y', 0) = f(y', 0).$$

Differentiating F, we obtain

$$D^\alpha F(x',x_n) = \sum_{i=1}^{k+1} A_i(-1/i)^{\alpha_n}(D^\alpha f)(x',-x_n/i)$$

if $x \in K_a^-$. Therefore,

$$\lim_{\substack{x\to(y',0),\\ x_n<0}} D^\alpha F(x',x_n) = \sum_{i=1}^{k+1} A_i(-1/i)^{\alpha_n} D^\alpha f(y',0) = D^\alpha f(y',0)$$

from system (6.3) ($s = \alpha_n \leqslant k$). We have shown that $F \in C^k(\overline{K_a})$. Moreover, the existence of a constant $M > 0$ such that the inequalities

$$\|F\|_{C^k(\overline{K_a})} \leqslant M\|f\|_{C^k(\overline{K_a^+})}, \quad \|F\|_{H^k(K_a)} \leqslant M\|f\|_{H^k(K_a^+)} \tag{6.4}$$

are valid for all $f \in C^k(\overline{K_a^+})$ follows from the definition of F and property 5) of Sobolev spaces.

Now suppose that $f \in H^k(K_a^+)$. By Lemma 6.1, there exists a sequence $f_m \in C^\infty(\overline{K_a^+})$ such that $f_m \to f$, $m \to \infty$, in $H^k(K_a^+)$. Let $F_m \in C^k(\overline{K_a})$ be the corresponding extensions constructed above. The second estimate in (6.4) means that

$$\|F_m - F_p\|_{H^k(K_a)} \leqslant M\|f_m - f_p\|_{H^k(K_a^+)} \to 0, \quad m,p \to \infty,$$

i. e., the sequence F_m is a Cauchy sequence in $H^k(K_a)$. This means that F_m converges in $H^k(K_a)$ to the function $F \in H^k(K_a)$. But then $F_m|_{K_a^+} = f_m \to F|_{K_a^+}$ in $H^k(K_a^+)$ implying $F|_{K_a^+} = f$, i. e., F is an extension of f. Passing to the limit as $m \to \infty$ in the inequality

$$\|F_m\|_{H^k(K_a)} \leqslant M\|f_m\|_{H^k(K_a^+)},$$

we extend the corresponding estimate to the whole space $H^k(K_a^+)$. □

Note that in Lemma 6.2 a *bounded extension operator* is constructed, acting from $C^k(\overline{K_a^+})$ to $C^k(\overline{K_a})$ (from $H^k(K_a^+)$ to $H^k(K_a)$).

Theorem 6.2. *Let Q and Ω be bounded domains in $\mathbb{R}^n$ such that $\partial Q \in C^k$ and $\overline{Q} \subset \Omega$. Then each function $f \in C^k(\overline{Q})$ can be extended to Ω by a compactly supported function $F \in C_0^k(\Omega)$, and the inequality*

$$\|F\|_{C^k(\overline{\Omega})} \leqslant M\|f\|_{C^k(\overline{Q})}$$

holds with a constant $M > 0$ independent of f. The same statement is true when replacing the space C^k with H^k.

Proof. For an arbitrary point $x^* \in \partial Q$, consider its neighborhood U_{x^*}, $\overline{U}_{x^*} \subset \Omega$ and a C^k-diffeomorphism $y = y(x)$ of this neighborhood onto the neighborhood V_{x^*} of the origin such that

$$U_{x^*} \cap \partial Q \mapsto V_{x^*} \cap \{y_n = 0\}, \quad U_{x^*} \cap Q \mapsto V_{x^*} \cap \{y_n > 0\}.$$

Taking some cube K_a, $\overline{K_a} \subset V_{x^*}$, we denote by W_{x^*} its preimage under the indicated diffeomorphism, $\overline{W_{x^*}} \subset U_{x^*}$.

The transition to new coordinates $y = y(x)$ that transfers part of the boundary $\partial Q \cap U_{x^*}$ to the hyperplane $\{y_n = 0\}$ is called the *local straightening of the boundary.*

If $x^* \in Q$, then as $W_{x^*} \subset Q$ we take an arbitrary neighborhood of the point x^* entirely lying in Q and do not consider any diffeomorphism $y = y(x)$.

Thus, a covering of $\overline{Q}$ by open sets W_{x^*}, $x^* \in \overline{Q}$, has been constructed. Let $\{\varphi_j\}_{j=1}^N$ be a partition of unity subject to this cover. If W_j is a cover element containing $\operatorname{supp}\varphi_j$, then we denote by T_j the corresponding variable transformation operator $f(x) \mapsto g(y) = f(x(y))$, $x \in W_j$ (we assume T_j to be the identity operator in the case of $W_j \subset Q$). It is clear that T_j is a linear homeomorphism of the Banach spaces $C^k(\overline{W_j})$ and $C^k(\overline{K_a})$, as well as $C^k(\overline{W_j \cap Q})$ and $C^k(\overline{K_a^+})$. Next, along with the functions φ_j, we introduce cut-off functions $\psi_j \in C_0^\infty(W_j)$ such that $\psi_j(x) = 1$ on the set $\operatorname{supp}\varphi_j$, $j = 1, \ldots, N$. Finally, let $C_j : C^k(\overline{K_a^+}) \to C^k(\overline{K_a})$ be the bounded linear extension operator constructed in Lemma 6.2.

The desired extension of an arbitrary function $f \in C^k(\overline{Q})$ is given by the formula

$$F = \sum_{j=1}^{N} \psi_j T_j^{-1} C_j T_j(\varphi_j f) \tag{6.5}$$

(we assume that C_j is applied only when W_j intersects the boundary). In detail, multiplication by φ_j in (6.5) gives us a smooth function on the set $\overline{W_j \cap Q}$, where the corresponding change of variables T_j is defined, transforming this function into a function from $C^k(\overline{K_a^+})$. The latter is smoothly extended into K_a by the operator C_j described above. The resulting extension is then carried back to W_j. Multiplication by ψ_j completes the procedure, defining a compactly supported function in W_j. It is clear that $F \in C_0^k(\Omega)$. On the other hand, the transformation C_j is dropped in (6.5) if $x \in \overline{Q}$, ensuring the equality

$$F(x) = \sum_{j=1}^{N} \psi_j(x)\varphi_j(x) f(x) = \sum_{j=1}^{N} \varphi_j(x) f(x) = f(x) \quad (x \in \overline{Q}).$$

Lemma 6.2 also guarantees the boundedness of the linear extension operator from $H^k(K_a^+)$ to $H^k(K_a)$. We use the previous notation C_j for it. In addition, the variable transformation operators T_j act as linear homeomorphisms in the corresponding Sobolev spaces. Therefore, formula (6.5) can be used to construct a bounded extension operator in the case of Sobolev spaces. This operator converts a function $f \in H^k(Q)$ into a compactly supported function $F \in H^k(\Omega)$, $\operatorname{supp} F \subset \Omega$, and the inequalities

$$\|F\|_{C^k(\overline{\Omega})} \leqslant M\|f\|_{C^k(\overline{Q})}, \quad \|F\|_{H^k(\Omega)} \leqslant M\|f\|_{H^k(Q)}$$

are a consequence of the boundedness of the operators T_j, T_j^{-1}, C_j and multiplication by φ_j and ψ_j in the corresponding spaces. □

Remark 6.1. Note that the bounded linear extension operator from $H^k(Q)$ to $H^k(\Omega)$ can be constructed for a much wider class of bounded domains with *Lipschitz boundary*. The proof of this fact is beyond the scope of the textbook; see [19, Chapter 4, Section 3].

Theorem 6.3. *Let Q be a bounded domain in $\mathbb{R}^n$ and $\partial Q \in C^k$. Then the set $C^\infty(\overline{Q})$ is dense in the space $H^k(Q)$.*

Proof. Let us choose a bounded domain Ω containing Q together with its closure, $\overline{Q} \subset \Omega$. Taking an arbitrary function $f \in H^k(Q)$, consider its compactly supported extension $F \in H^k(\Omega)$, the existence of which is guaranteed by Theorem 6.2. If $\{F_h\}_{h>0}$ is the averaging of F in Ω then, for small values of h, the functions F_h belong to the space $C_0^\infty(\Omega)$ and $\|F - F_h\|_{H^k(\Omega)} \to 0$ as $h \to 0$ due to property 4) of Sobolev spaces. Then $F_h|_Q \in C^\infty(\overline{Q})$ and $\|f - F_h\|_{H^k(Q)} \leqslant \|F - F_h\|_{H^k(\Omega)} \to 0$ as $h \to 0$. □

Separability of $H^k(Q)$. Compactness of the embedding of $H^1(Q)$ into $L_2(Q)$

Theorem 6.4. *Let Q be a bounded domain in $\mathbb{R}^n$ and $\partial Q \in C^k$. Then the space $H^k(Q)$ is separable.*

Proof. We can assume without loss of generality that

$$\overline{Q} \subset K_\pi = \{x \in \mathbb{R}^n : |x_i| < \pi,\ i = 1, \dots, n\}.$$

Take an arbitrary function $v \in H^k(Q)$. Let $V \in H^k(K_\pi)$ be its compactly supported extension guaranteed by Theorem 6.2. Based on property 4) of Sobolev spaces, we approximate the function V in $H^k(K_\pi)$ by functions from $C_0^\infty(K_\pi)$. At the same time, any $C_0^\infty(K_\pi)$-function can be approximated in $H^k(K_\pi)$ by partial sums of its Fourier series. Let us elaborate on this point.

It is well known that the functions

$$\{(2\pi)^{-n/2} e_m(x) = (2\pi)^{-n/2} e^{i(m,x)} : m \in \mathbb{Z}^n\}$$

form an orthonormal basis in the Hilbert space $L_2(K_\pi)$. Here, i denotes the imaginary unit and $(m, x) = m_1 x_1 + \cdots + m_n x_n$. We use the usual notation $|m| = (m_1^2 + \cdots + m_n^2)^{1/2}$ and $m^\alpha = m_1^{\alpha_1} \cdot \dots \cdot m_n^{\alpha_n}$, where α is a multiindex.

Any function $V \in L_2(K_\pi)$ can be expanded into the Fourier series:

$$V(x) = (2\pi)^{-n/2} \sum_{m \in \mathbb{Z}^n} V_m e^{i(m,x)}, \quad V_m = (2\pi)^{-n/2} \int_{K_\pi} V(x) e^{-i(m,x)}\, dx,$$

and

$$\Big\| V - (2\pi)^{-n/2} \sum_{|m| \leqslant N} V_m e_m \Big\|_{L_2(K_\pi)} \to 0 \quad \text{as } N \to \infty.$$

If $V \in C_0^\infty(K_\pi)$, then the corresponding Fourier series can be given for all partial derivatives $D^\alpha V$,

$$D^\alpha V(x) = (2\pi)^{-n/2} \sum_{m \in \mathbb{Z}^n} (im)^\alpha V_m e^{i(m,x)}$$

since

$$(2\pi)^{-n/2} \int\limits_{K_\pi} D^\alpha V(x) e^{-i(m,x)} dx = (im)^\alpha V_m$$

(the last equality is obtained by repeated integration by parts). Hence

$$\Big\| D^\alpha V - (2\pi)^{-n/2} \sum_{|m| \leqslant N} (im)^\alpha V_m e_m \Big\|_{L_2(K_\pi)} \to 0 \quad \text{as } N \to \infty.$$

But this implies

$$\Big\| D^\alpha V - D^\alpha \Big((2\pi)^{-n/2} \sum_{|m| \leqslant N} V_m e_m \Big) \Big\|_{L_2(K_\pi)} \to 0 \quad \text{as } N \to \infty.$$

In other words, the Fourier series of a smooth compactly supported function converges to this function in $H^k(K_\pi)$ (for all $k = 0, 1, \ldots$) and, therefore, any $C_0^\infty(K_\pi)$-function is approximated in $H^k(K_\pi)$ by trigonometric polynomials. It is also clear that the approximation can be made by polynomials with only rational coefficients. As a result, we obtain that any function from $H^k(Q)$ can be approximated as accurately as desired in the space $H^k(Q)$ by elements of a countable set of trigonometric polynomials with rational coefficients. □

Theorem 6.5. *Let Q be a bounded domain in $\mathbb{R}^n$ and $\partial Q \in C^1$. Then the space $H^1(Q)$ is compactly embedded in the space $L_2(Q)$.*

Proof. The theorem means that from any sequence $v_m \in H^1(Q)$, $m = 1, 2 \ldots$, bounded in $H^1(Q)$ one can extract a subsequence v_{m_p}, $p = 1, 2, \ldots$, being a Cauchy sequence in $L_2(Q)$. Let us show it. Take a bounded domain $\Omega \supset \overline{Q}$. Based on Theorem 6.2, we move on to the sequence of extensions $V_m(x)$ with compact supports in Ω, bounded in $H^1(\Omega)$, $\|V_m\|_{H^1(\Omega)} \leqslant C$. Taking into account property 4) of Sobolev spaces, we can assume that these functions V_m belong to $C_0^\infty(\Omega)$ (and are extended by zero to $\mathbb{R}^n \setminus \Omega$).

All this allows us to apply the Fourier transform in $\mathbb{R}^n$,

$$V_m(x) = (2\pi)^{-n/2} \int\limits_{\mathbb{R}^n} \widehat{V}_m(\xi) e^{i(\xi,x)} d\xi, \quad \widehat{V}_m(\xi) = (2\pi)^{-n/2} \int\limits_{\mathbb{R}^n} V_m(x) e^{-i(\xi,x)} dx$$

$$((\xi, x) = \xi_1 x_1 + \cdots + \xi_n x_n).$$

Due to (2.6), $\widehat{D^\alpha V_m}(\xi) = (i\xi)^\alpha \widehat{V}_m(\xi)$ for $V_m \in C_0^\infty(\mathbb{R}^n)$. Moreover, the Plancherel theorem states that $\|V_m\|_{L_2(\mathbb{R}^n)} = \|\widehat{V}_m\|_{L_2(\mathbb{R}^n)}$.

Being bounded in the Hilbert space $L_2(\Omega)$, the sequence $V_m(x)$ contains a weakly convergent subsequence $V_{m_p}(x)$. Note that the value of the function $\widehat{V}_m(\xi)$ at a fixed point $\xi \in \mathbb{R}^n$ can be interpreted as the inner product

$$\widehat{V}_{m_p}(\xi) = (V_{m_p}, \varphi_\xi)_{L_2(\Omega)}, \quad \varphi_\xi = (2\pi)^{-n/2} e^{i(\xi,x)}.$$

Therefore, the sequence $\widehat{V}_{m_p}(\xi)$ converges pointwise in $\mathbb{R}^n$. Moreover, the Cauchy–Schwarz inequality gives

$$|\widehat{V}_{m_p}(\xi)| \leqslant (2\pi)^{-n/2}|\Omega|^{1/2}C,$$

i. e., the sequence $\widehat{V}_{m_p}(\xi)$ is uniformly bounded. We took advantage of the boundedness of Ω in the last two formulas.

Let us prove that the sequence V_{m_p} is a Cauchy sequence in $L_2(\Omega)$. By the Plancherel theorem,

$$\|V_{m_p} - V_{m_q}\|^2_{L_2(\Omega)} = \|\widehat{V}_{m_p} - \widehat{V}_{m_q}\|^2_{L_2(\mathbb{R}^n)} = I_1 + I_2,$$

where

$$I_1 = \int\limits_{|\xi|>R} |\widehat{V}_{m_p}(\xi) - \widehat{V}_{m_q}(\xi)|^2\, d\xi, \quad I_2 = \int\limits_{|\xi|<R} |\widehat{V}_{m_p}(\xi) - \widehat{V}_{m_q}(\xi)|^2\, d\xi.$$

Let us first estimate I_1. Applying a number of obvious inequalities, the formula for the Fourier transform of derivatives and the Plancherel theorem, we obtain

$$\begin{aligned}
I_1 &= \int\limits_{|\xi|>R} |\widehat{V}_{m_p}(\xi) - \widehat{V}_{m_q}(\xi)|^2\, d\xi \leqslant \int\limits_{|\xi|>R} \frac{|\xi|^2}{R^2}|\widehat{V}_{m_p}(\xi) - \widehat{V}_{m_q}(\xi)|^2 d\xi \leqslant \\
&\leqslant \frac{1}{R^2}\int\limits_{\mathbb{R}^n} |\xi|^2|\widehat{V}_{m_p}(\xi) - \widehat{V}_{m_q}(\xi)|^2\, d\xi = \\
&= \frac{1}{R^2}\sum_{j=1}^n \int\limits_{\mathbb{R}^n} |(i\xi_j)(\widehat{V}_{m_p}(\xi) - \widehat{V}_{m_q}(\xi))|^2 d\xi = \\
&= \frac{1}{R^2}\sum_{j=1}^n \int\limits_{\mathbb{R}^n} |(V_{m_p}(x) - V_{m_q}(x))_{x_j}|^2 dx \leqslant \frac{1}{R^2}\|V_{m_p} - V_{m_q}\|^2_{H^1(\Omega)} \leqslant \frac{4C^2}{R^2},
\end{aligned}$$

thus the integral I_1 can be made arbitrarily small by choosing R large enough. Let us choose R such that $I_1 < \varepsilon^2/2$.

Taking into account the properties of $\widehat{V}_{m_p}(\xi)$ established above, we apply Lebesgue's dominated convergence theorem to I_2 to get $I_2 \to 0$, $p, q \to \infty$. Thus, $I_2 < \varepsilon^2/2$ if p and

q are large enough. We have shown that $\|V_{m_p} - V_{m_q}\|_{L_2(\Omega)} < \varepsilon$ for all sufficiently large numbers p and q. The theorem is now proven. □

Remark 6.2. The assumptions of Theorem 6.5 can be weakened. Condition $\partial Q \in C^1$ on the boundary of the domain can be replaced by other conditions, e. g., the *cone condition*; see [23]. The domain can also be unbounded, but must be "narrow" at infinity. Without dwelling on this issue in more detail here, we only note that the statement is not true in the case $Q = \mathbb{R}^n$.

Example 6.1. Let us check that the embedding of $H^1(\mathbb{R}^n)$ into $L_2(\mathbb{R}^n)$ is not compact. Consider a sequence $v_m(x) = c_0\omega_{1/4}(x - g_m)$, where $\omega_h(x)$ is the mollifier from Definition 3.1, the number $c_0 > 0$ is chosen such that $\|v_0\|_{H^1(\mathbb{R}^n)} = 1$, and $g_m = (m, 0, \dots, 0)$. Obviously,

$$\|v_m\|_{H^1(\mathbb{R}^n)} = \|v_0\|_{H^1(\mathbb{R}^n)} = 1 \quad \text{and} \quad \|v_m - v_k\|^2_{L_2(\mathbb{R}^n)} = 2\|v_0\|^2_{L_2(\mathbb{R}^n)} = 2$$

for all $k, m \in \mathbb{N}$. Therefore, the sequence $\{v_m\}$ does not contain convergent subsequences in $L_2(\mathbb{R}^n)$.

Trace theorem

Considering boundary-value problems in Sobolev spaces, we are dealing with the boundary values of the solution. But functions from Sobolev spaces are specified up to sets of measure zero. Therefore, the boundary values are not formally defined. To overcome these difficulties, the concept of the *trace* of a function is introduced in this section.

Let Q be a bounded domain in $\mathbb{R}^n$ and $\partial Q \in C^1$. Being a compact set, the boundary ∂Q can be represented as the union of a finite number of *simple pieces* S_l, $\partial Q = \bigcup_{l=1}^N S_l$, where

$$S_l = \{x \in \mathbb{R}^n : x_{i_l} = \varphi_l(x'),\ x' \in D_l\}, \quad x' = (x_1, \dots, x_{i_l-1}, x_{i_l+1}, \dots, x_n),$$

D_l is an $(n-1)$-dimensional domain, and $\varphi_l \in C^1(\overline{D}_l)$. Let us place Q in the cube K_a, then all D_l will lie in the corresponding $(n-1)$-dimensional cubes with side $2a$.

We assume that functions from $C^1(\overline{Q})$ are already extended to K_a by functions from $C_0^1(K_a)$. Therefore, we take an arbitrary function $u \in C_0^1(K_a)$ and consider its restriction $u|_{S_l}$ on S_l, $l = 1, \dots, N$. At this step, we assume for brevity that $i_l = n$ and $\varphi_l = \varphi$, thus $u|_{S_l} = u(x', \varphi(x'))$, $x' \in D_l$. Using the Newton–Leibniz formula, we write

$$u(x) = u(x', \varphi(x')) = \int_{-a}^{\varphi(x')} u_{x_n}(x', \xi)\, d\xi \quad (x \in S_l),$$

hence

$$|u(x)|^2 \leqslant 2a \int_{-a}^{a} |u_{x_n}(x',\xi)|^2 \, d\xi \quad (x \in S_l).$$

Multiplying the last inequality by $\sqrt{\varphi_{x_1}^2 + \cdots + \varphi_{x_{n-1}}^2 + 1}$ and integrating over D_l, we arrive at the inequality

$$\int_{D_l} |u(x',\varphi(x'))|^2 \sqrt{\varphi_{x_1}^2 + \cdots + \varphi_{x_{n-1}}^2 + 1}\, dx' \leqslant k_1 \int_{K_a} |u_{x_n}|^2 \, dx$$

or

$$\|u|_{S_l}\|^2_{L_2(S_l)} \leqslant k_1 \|u_{x_n}\|^2_{L_2(K_a)} \leqslant k_1 \|u\|^2_{H^1(K_a)} \leqslant k_2 \|u\|^2_{H^1(Q)}$$

with positive constants k_1, k_2 independent of $u \in C^1(\overline{Q})$ (they depend only on the function φ and on the extension operator from $H^1(Q)$ to $H^1(K_a)$). Summing over $l = 1, \ldots, N$, we come to the inequality

$$\|u|_{\partial Q}\|_{L_2(\partial Q)} \leqslant k_3 \|u\|_{H^1(Q)} \tag{6.6}$$

on the set $C^1(\overline{Q})$.

Next, the obtained estimate is extended in a standard way from the set $C^1(\overline{Q})$ dense in $H^1(Q)$ to the whole space $H^1(Q)$. The arguments below show in which sense the boundary values of an $H^1(Q)$-function should be understood.

By Theorem 6.3, any function $u \in H^1(Q)$ can be approximated in $H^1(Q)$ by a sequence of smooth functions u_m,

$$\|u - u_m\|_{H^1(Q)} \to 0, \quad m \to \infty, \quad u_m \in C^1(\overline{Q}), \quad m = 1, 2, \ldots .$$

Substituting $u_m - u_p$ for u in (6.6), we get $\|u_m|_{\partial Q} - u_p|_{\partial Q}\|_{L_2(\partial Q)} \to 0$, $m, p \to \infty$, i. e., the sequence $u_m|_{\partial Q}$ is a Cauchy sequence in $L_2(\partial Q)$. The limit function is denoted by $u|_{\partial Q}$ and is called the *trace* of the function $u \in H^1(Q)$ on the boundary. The independence of the trace $u|_{\partial Q}$ from the choice of an approximating sequence can easily be verified. Using u_m as u again in (6.6) and passing to the limit as $m \to \infty$, we obtain the corresponding estimate on the whole space $H^1(Q)$. Let us summarize the performed steps.

Theorem 6.6. *Let $Q \subset \mathbb{R}^n$ be a bounded domain with C^1-boundary. Then, for any function $u \in H^1(Q)$, its trace $u|_{\partial Q}$ on the boundary is defined. The trace operator is bounded from $H^1(Q)$ to $L_2(\partial Q)$.*

Remark 6.3. The trace of a function from $H^1(Q)$ is defined as an element of the space $L_2(\partial Q)$ but in fact belongs to a narrower space. This space is usually denoted by $H^{1/2}(\partial Q)$ and is called the *trace space*. By definition, the trace space consists of all functions on the boundary that can be extended into the domain by functions from $H^1(Q)$. Without giving an explicit description of the trace space, $H^{1/2}(\partial Q)$ is provided with the so-called *infimum norm*, assuming

$$\|\gamma\|_{H^{1/2}(\partial Q)} = \inf_{\substack{u\in H^1(Q),\\ u|_{\partial Q}=\gamma}} \|u\|_{H^1(Q)}. \tag{6.7}$$

Then the trace operator immediately becomes bounded from $H^1(Q)$ to the trace space $H^{1/2}(\partial Q)$ with norm equal to 1.

It is easy to verify that the trace space is a Banach space with respect to norm (6.7). Without going into detail, we note that the topology of the trace space is stronger than the L_2-topology and, as an operator from $H^1(Q)$ to $L_2(\partial Q)$, the trace operator is compact.

In fact, the trace space has the structure of a Hilbert space inducing an equivalent norm. Moreover, it can be described explicitly. In a certain sense, $H^{1/2}(\partial Q)$ consists of the functions from $L_2(\partial Q)$ that have fractional derivatives of order 1/2, also belonging to $L_2(\partial Q)$. In this book, we do not prove these statements but refer the reader, e. g., to [9] (see also exercises 9–11).

Remark 6.4. Theorem 6.6 considers traces on the boundary of a domain. The result is obviously valid for any $(n-1)$-dimensional surface $S \subset \overline{Q}$ of class C^1, representable as the union of a finite number of simple pieces.

Space $\mathring{H}^1(Q)$. Friedrichs's inequality

By adding to the set $C^\infty(\overline{Q})$ the limits of all sequences of functions from $C^\infty(\overline{Q})$ converging with respect to the $H^1(Q)$-norm, we obtain the entire space $H^1(Q)$. This is no longer the case if we limit ourselves to compactly supported functions.

Definition 6.2. The Hilbert space $\mathring{H}^1(Q)$ is the closure of the set $C_0^\infty(Q)$ in $H^1(Q)$.

If $u \in \mathring{H}^1(Q)$ and $\partial Q \in C^1$, then $u|_{\partial Q} = 0$. Indeed, any function $u \in \mathring{H}^1(Q)$ is the limit of a sequence $u_m \in C_0^\infty(Q)$ converging in $H^1(Q)$. By Theorem 6.6, we have $\|u|_{\partial Q} - u_m|_{\partial Q}\|_{L_2(\partial Q)} \leqslant c\,\|u - u_m\|_{H^1(Q)} \to 0$, $m \to \infty$, and it only remains to note that $u_m|_{\partial Q} = 0$ for all m. The inverse implication $u|_{\partial Q} = 0 \Rightarrow u \in \mathring{H}^1(Q)$ is also valid, but not as obvious.

Theorem 6.7. *Let $Q \subset \mathbb{R}^n$ be a bounded domain and $\partial Q \in C^1$. Then $\mathring{H}^1(Q) = \{u \in H^1(Q) : u|_{\partial Q} = 0\}$.*

To prove Theorem 6.7, we need the following auxiliary statement.

Lemma 6.3. *Suppose that $u \in H^1(K_a^+)$ and $u|_{x_n=0} = 0$, then*

$$\|u\|_{L_2(K_{a,\delta}^+)} = o(\delta) \quad as\ \delta \to 0, \tag{6.8}$$

where $K_{a,\delta}^+ = \{x \in K_a^+ : 0 < x_n < \delta\}$.

Proof. For any function $u \in C^1(\overline{K_a^+})$ and $0 < \rho < \delta$, we can obviously write

$$u(x', \rho) - u(x', 0) = \int_0^\rho u_{x_n}(x', x_n)\, dx_n.$$

Therefore,

$$|u(x', \rho) - u(x', 0)|^2 \leqslant \rho \int_0^\rho |u_{x_n}(x', x_n)|^2\, dx_n.$$

Integrating this inequality with respect to x', we obtain

$$\|u|_{x_n=\rho} - u|_{x_n=0}\|^2_{L_2(K_a')} \leqslant \rho \|u\|^2_{H^1(K^+_{a,\rho})} \leqslant \rho \|u\|^2_{H^1(K^+_{a,\delta})}, \tag{6.9}$$

where $K_a' = \{x' \in \mathbb{R}^{n-1} : |x_i| < a,\ i = 1, \dots, n-1\}$.

By Lemma 6.1, the set $C^\infty(\overline{K_a^+})$ is dense in $H^1(K_a^+)$, thus inequality (6.9) holds for all $u \in H^1(K_a^+)$. It expresses the L_2-continuity of the trace under shifts of the manifold.

If $u|_{x_n=0} = 0$, then integration of (6.9) over $\rho \in (0, \delta)$ gives

$$\|u\|^2_{L_2(K^+_{a,\delta})} \leqslant \frac{\delta^2}{2} \|u\|^2_{H^1(K^+_{a,\delta})},$$

thus relation (6.8) follows from the absolute continuity of the Lebesgue integral. □

Remark 6.5. In fact, it is important to make sure that integrating the (squared) L_2-norm of the trace of a function from the Sobolev space, we obtain the (squared) L_2-norm of this function,

$$\int_0^\delta \|u|_{x_n=\rho}\|^2_{L_2(K_a')}\, d\rho = \|u\|^2_{L_2(K^+_{a,\delta})}.$$

To prove this, we approximate the function $u \in H^1(K_a^+)$ by a sequence of smooth functions,

$$\{u_j\} \subset C^\infty(\overline{K_a^+}), \quad \|u - u_j\|_{H^1(K_a^+)} \to 0.$$

By definition of the trace,

$$\|u|_{x_n=\rho}\|^2_{L_2(K_a')} = \lim_{j\to\infty} \|u_j(\cdot, \rho)\|^2_{L_2(K_a')}.$$

Moreover, it follows from inequality (6.9) that the norms $\|u_j|_{x_n=\rho}\|_{L_2(K_a')}$ are uniformly bounded, i. e., the inequality $\|u_j|_{x_n=\rho}\|_{L_2(K_a')} \leqslant c$ holds with a positive constant c independent of j and ρ. The reasoning concludes with the use of Fubini's theorem,

$$\int_0^\delta \|u_j(\cdot,\rho)\|^2_{L_2(K'_a)}\,d\rho = \|u_j\|^2_{L_2(K^+_{a,\delta})},$$

and the application of Lebesgue's dominated convergence theorem to the resulting equality.

Proof of Theorem 6.7. We must show that

$$\{u \in H^1(Q) : u|_{\partial Q} = 0\} \subset \mathring{H}^1(Q).$$

Using local boundary straightening and partition of unity (see the proof of Theorem 6.2), we reduce the verification of this relation to the following statement.

For any function $u \in H^1(K^+_a)$ satisfying the condition $u|_{x_n=0} = 0$, there exists a sequence $\{v_j\} \subset C^\infty(\overline{K^+_a})$ such that

$$v_j(x) = 0 \quad (x \in K^+_{a,\sigma},\ \sigma = \sigma(j))$$

and $\|u - v_j\|_{H^1(K^+_a)} \to 0$ as $j \to \infty$.

Let $u \in H^1(K^+_a)$ and $u|_{x_n=0} = 0$. By virtue of Lemma 6.3 and the absolute continuity of the Lebesgue integral, for any $\varepsilon > 0$ there exists a number $\delta = \delta(\varepsilon)$, $0 < \delta < \min\{1, a/2\}$, such that

$$\|u\|_{L_2(K^+_{a,\delta})} < \varepsilon\delta \quad \text{and} \quad \|u\|_{H^1(K^+_{a,\delta})} < \varepsilon. \tag{6.10}$$

By Lemma 6.1, there exists a sequence $\{u_j\} \subset C^\infty(\overline{K^+_a})$ converging to u in $H^1(K^+_a)$. Therefore, we can choose a number $N = N(\varepsilon)$ such that the inequality

$$\|u - u_N\|_{H^1(K^+_a)} < \varepsilon\delta \tag{6.11}$$

holds.

Since $0 < \delta < 1$, we deduce from inequalities (6.10) and (6.11) that

$$\|u - u_N\|_{H^1(K^+_a)} < \varepsilon, \tag{6.12}$$

$$\|u_N\|_{L_2(K^+_{a,\delta})} \leqslant \|u_N - u\|_{L_2(K^+_a)} + \|u\|_{L_2(K^+_{a,\delta})} < 2\varepsilon\delta, \tag{6.13}$$

$$\|u_N\|_{H^1(K^+_{a,\delta})} \leqslant \|u_N - u\|_{H^1(K^+_a)} + \|u\|_{H^1(K^+_{a,\delta})} < 2\varepsilon. \tag{6.14}$$

Let us introduce the cut-off function

$$\xi_\delta(x_n) = \int_{\delta/2}^\infty \omega_{\delta/3}(x_n - \tau)\,d\tau \quad (x_n \in \mathbb{R}),$$

where $\omega_{\delta/3}(x_n - \tau)$ is the mollifier. It follows from the definition of the mollifier that $\xi_\delta \in C^\infty(\mathbb{R})$, $0 \leqslant \xi_\delta(x_n) \leqslant 1$ $(x_n \in \mathbb{R})$, $\xi_\delta(x_n) = 1$ for $x_n \geqslant 5\delta/6$, $\xi_\delta(x_n) = 0$ for $x_n \leqslant \delta/6$, and $|\xi'_\delta(x_n)| \leqslant k_1/\delta$, where $k_1 > 0$ does not depend on δ.

Based on the properties of the function $\xi_\delta(x_n)$ and inequalities (6.12)–(6.14), we obtain

$$\begin{aligned}
\|u_N - \xi_\delta u_N\|_{H^1(K_a^+)} &= \|u_N - \xi_\delta u_N\|_{H^1(K_{a,\delta}^+)} \leqslant \\
&\leqslant \left(\|u_N(1-\xi_\delta)\|^2_{L_2(K_{a,\delta}^+)} + \big\||\nabla u_N|(1-\xi_\delta) + |u_N\xi'_\delta|\big\|^2_{L_2(K_{a,\delta}^+)}\right)^{1/2} \leqslant \\
&\leqslant \left(\|u_N\|^2_{L_2(K_{a,\delta}^+)} + 2\||\nabla u_N|\|^2_{L_2(K_{a,\delta}^+)} + 2\|u_N\xi'_\delta\|^2_{L_2(K_{a,\delta}^+)}\right)^{1/2} \leqslant \\
&\leqslant \left(4\varepsilon^2\delta^2 + 8\varepsilon^2 + \frac{2k_1^2}{\delta^2}4\varepsilon^2\delta^2\right)^{1/2} \leqslant k_2\varepsilon,
\end{aligned}$$

with a constant $k_2 > 0$ independent of ε.

Finally, we have $u_N\xi_\delta \in C^\infty(\overline{K_a^+})$, $u_N(x)\xi_\delta(x_n) = 0$ $(x \in K^+_{a,\delta/6})$ and

$$\|u - \xi_\delta u_N\|_{H^1(K_a^+)} \leqslant \|u - u_N\|_{H^1(K_a^+)} + \|u_N - \xi_\delta u_N\|_{H^1(K_a^+)} \leqslant (1 + k_2)\varepsilon. \qquad \square$$

Theorem 6.8. *Let Q be a bounded domain in $\mathbb{R}^n$ and $d = \operatorname{diam} Q$. Then for any function $u \in \mathring{H}^1(Q)$ the following inequality holds:*

$$\int_Q |u|^2\,dx \leqslant d^2 \int_Q |\nabla u|^2\,dx. \tag{6.15}$$

Inequality (6.15) is called the *Friedrichs inequality*. Sometimes it may be referred to as the *Steklov inequality*.

Proof. It is enough to prove estimate (6.15) for $u \in C_0^\infty(Q)$. Without loss of generality, we can assume that $Q \subset (0,d)^n$. Using the Newton–Leibniz formula,

$$u(x', x_n) = \int_0^{x_n} u_\tau(x', \tau)\,d\tau,$$

and the Cauchy–Schwarz inequality, we have

$$|u(x', x_n)|^2 \leqslant d\int_0^d |u_\tau(x', \tau)|^2\,d\tau, \quad (x', x_n) \in (0,d)^n.$$

It remains to integrate the resulting inequality over $(0,d)^n$:

$$\int_Q |u|^2\,dx = \int_{(0,d)^n} |u|^2\,dx \leqslant d^2 \int_{(0,d)^n} |u_{x_n}(x)|^2\,dx \leqslant d^2 \int_Q |\nabla u(x)|^2\,dx. \qquad \square$$

Definition 6.3. Inner products $(\cdot,\cdot)_H$ and $(\cdot,\cdot)'_H$ in a Hilbert space H are said to be equivalent if there exist positive constants c_1 and c_2 such that the inequalities

$$c_1\|u\|_H \leqslant \|u\|'_H \leqslant c_2\|u\|_H$$

are valid for all $u \in H$.

Corollary 6.1. *The form $(u,v)'_{\mathring{H}^1(Q)} = (\nabla u, \nabla v)_{L_2^n(Q)}$ defines the inner product equivalent to the standard inner product $(u,v)_{\mathring{H}^1(Q)} = (\nabla u, \nabla v)_{L_2^n(Q)} + (u,v)_{L_2(Q)}$ in the space $\mathring{H}^1(Q)$.*

Integration by parts formula

Theorem 6.9. *Let Q be a bounded domain in $\mathbb{R}^n$, $\partial Q \in C^1$ and $\nu = \nu(x)$ be the unit outward normal vector to the boundary. Then the equality*

$$\int_Q u_{x_j} v\,dx = \int_{\partial Q} u|_{\partial Q}\, v|_{\partial Q}\, \nu_j\,dS - \int_Q u v_{x_j}\,dx \tag{6.16}$$

holds for all functions $u, v \in H^1(Q)$.

Proof. If $u \in C^1(\overline{Q})$ and $v \in C^1(\overline{Q})$, then relation (6.16) is the well-known formula for integration by parts. It can easily be transferred to the entire Sobolev space $H^1(Q)$ due to the density of the set $C^1(\overline{Q})$ in $H^1(Q)$ and the trace theorem. Indeed, if sequences u_m and v_m from $C^1(\overline{Q})$ approximate functions u and v in $H^1(Q)$, then $u_m|_{\partial Q}$ and $v_m|_{\partial Q}$ approximate traces $u|_{\partial Q}$ and $v|_{\partial Q}$ of these functions on the boundary with respect to the norm of the space $L_2(\partial Q)$. Therefore, in equality

$$\int_Q (u_m)_{x_j} v_p\,dx = \int_{\partial Q} u_m|_{\partial Q}\, v_p|_{\partial Q}\, \nu_j\,dS - \int_Q u_m (v_p)_{x_j}\,dx$$

one can pass to the limit sequentially as $m \to \infty$ and $p \to \infty$ to obtain the required relation for u and v. We emphasize that since we are dealing with functions from $H^1(Q)$, the quantities $u|_{\partial Q}$ and $v|_{\partial Q}$ denote traces of functions on the boundary in the sense of this section. □

Sobolev embedding theorem

The following theorem establishes a connection between Sobolev spaces and spaces of continuously-differentiable functions. This connection is extremely important in analysis. In particular, this theorem allows one to obtain sufficient conditions under which weak solutions of boundary-value problems are simultaneously classical ones.

Theorem 6.10. *Let Q be a bounded domain in $\mathbb{R}^n$ with $C^{k+[n/2]+1}$-boundary. Then the Sobolev space $H^{k+[n/2]+1}(Q)$ is continuously embedded in the space $C^k(\overline{Q})$. This means that any function $u \in H^{k+[n/2]+1}(Q)$ coincides a. e. with a k-times continuously-differentiable function in $\overline{Q}$, also denoted by u, and*

$$\|u\|_{C^k(\overline{Q})} \leqslant c\,\|u\|_{H^{k+[n/2]+1}(Q)} \tag{6.17}$$

for all $u \in H^{k+[n/2]+1}(Q)$, where $c = c(Q,k) > 0$ is a constant.

Proof. Let us choose an arbitrary bounded domain Ω containing $\overline{Q}$. First, we prove estimate (6.17) (with Q replaced by Ω) for functions $U \in C_0^\infty(\Omega)$. For this, we use the Fourier transform

$$\widehat{U}(\xi) = (2\pi)^{-n/2} \int_\Omega e^{-i(\xi,x)} U(x)dx \quad (\xi \in \mathbb{R}^n),$$

whence

$$U(x) = (2\pi)^{-n/2} \int_{\mathbb{R}^n} e^{i(\xi,x)} \widehat{U}(\xi)d\xi$$

and, therefore,

$$D^\alpha U(x) = (2\pi)^{-n/2} \int_{\mathbb{R}^n} (i\xi)^\alpha e^{i(\xi,x)} \widehat{U}(\xi)d\xi$$

$$((\xi,x) = \xi_1 x_1 + \cdots + \xi_n x_n,\ (i\xi)^\alpha = (i\xi_1)^{\alpha_1}\ldots(i\xi_n)^{\alpha_n}).$$

It should be noted here that $\widehat{U}(\xi)$ is an infinitely-differentiable rapidly decreasing function.

Due to obvious inequalities,

$$|(i\xi)^\alpha|^2 = \xi_1^{2\alpha_1}\ldots\xi_n^{2\alpha_n} \leqslant (\xi_1^2 + \cdots + \xi_n^2)^{|\alpha|} \leqslant (1 + |\xi|^2)^{|\alpha|},$$

we see that

$$|D^\alpha U(x)| \leqslant (2\pi)^{-n/2} \int_{\mathbb{R}^n} (1 + |\xi|^2)^{k/2} |\widehat{U}(\xi)|\, d\xi =$$

$$= (2\pi)^{-n/2} \int_{\mathbb{R}^n} (1 + |\xi|^2)^{-\frac{[n/2]+1}{2}} (1 + |\xi|^2)^{\frac{k+[n/2]+1}{2}} |\widehat{U}(\xi)| d\xi$$

for $|\alpha| \leqslant k$. The function $(1 + |\xi|^2)^{-[n/2]-1}$ is integrable over $\mathbb{R}^n$.

Indeed, if $n = 2m$, then $-2([n/2]+1)+n-1 = -2m-2+2m-1 = -3 < -1$. If $n = 2m+1$, then again $-2([n/2] + 1) + n - 1 = -2m - 2 + 2m + 1 - 1 = -2 < -1$.

Therefore, using the Cauchy–Schwarz inequality, we can estimate

$$|D^{\alpha}U(x)| \leqslant c_1\Big(\int_{\mathbb{R}^n}(1+|\xi|^2)^{k+[n/2]+1}|\widehat{U}(\xi)|^2\,d\xi\Big)^{1/2} \leqslant c_2\|U\|_{H^{k+[n/2]+1}(\Omega)}.$$

The last inequality follows from the Plancherel theorem, the formula for the Fourier images of derivatives and the algebraic inequality

$$(1+|\xi|^2)^l = (1+\xi_1^2+\cdots+\xi_n^2)^l \leqslant c_3(1+\xi_1^{2l}+\cdots+\xi_n^{2l}), \quad l = k+[n/2]+1.$$

Therefore,

$$\|U\|_{C^k(\overline{\Omega})} \leqslant c_4\,\|U\|_{H^l(\Omega)} \quad (U \in C_0^{\infty}(\Omega)). \tag{6.18}$$

Next, given a function $u \in H^l(Q)$, consider its compactly supported extension U to Ω,

$$U \in H^l(\Omega), \quad \operatorname{supp} U \subset \Omega, \quad \|U\|_{H^l(\Omega)} \leqslant c_5\|u\|_{H^l(Q)}.$$

Using the averaging procedure, we construct a sequence of compactly supported infinitely-differentiable functions converging to U in $H^l(\Omega)$,

$$U_m \in C_0^{\infty}(\Omega), \quad \|U - U_m\|_{H^l(\Omega)} \to 0, \quad m \to \infty.$$

It follows from (6.18) that $\|U_m - U_p\|_{C^k(\overline{\Omega})} \to 0$, $m, p \to \infty$. Since $C^k(\overline{\Omega})$ is a Banach space, there exists a function $U^* \in C^k(\overline{\Omega})$ for which $U_m \to U^*$ in $C^k(\overline{\Omega})$ and, therefore, in $H^k(\Omega)$, $m \to \infty$. On the other hand, $U_m \to U$ in $H^l(\Omega)$, and, therefore, in $H^k(\Omega)$, $m \to \infty$. Thus, $U(x) = U^*(x)$ a. e. in Ω. This proves that $U(x)$ coincides a. e. with the function from $C^k(\overline{\Omega})$. For the latter, we keep the notation $U(x)$. Passing to the limit as $m \to \infty$ in estimate (6.18) for U_m instead of U, we obtain the required inequality for the function U,

$$\|U\|_{C^k(\overline{\Omega})} \leqslant c_4\,\|U\|_{H^l(\Omega)} \leqslant c_4c_5\|u\|_{H^l(Q)} \quad (u \in H^l(Q)).$$

Finally, we have $u = U|_{\overline{Q}} \in C^k(\overline{Q})$ and

$$\|u\|_{C^k(\overline{Q})} \leqslant \|U\|_{C^k(\overline{\Omega})} \leqslant c_4c_5\|u\|_{H^l(Q)},$$

i. e., estimate (6.17) is valid. □

Remark 6.6. There is a version of Theorem 6.10 that does not use the condition $\partial Q \in C^{k+[n/2]+1}$. In this case, the internal C^k-smoothness of functions from the Sobolev space $H^l(Q)$ with the estimate

$$\|u\|_{C^k(\overline{Q_\delta})} \leqslant c\,(Q,\delta,k)\,\|u\|_{H^l(Q)} \quad (\delta > 0)$$

are guaranteed instead of (6.17). We leave the details to the reader.

Example 6.2. Let us take $n = 1$, $k = 0$ and $Q = (0,1)$. Theorem 6.10 guarantees the continuous embedding $H^1(0,1) \subset C[0,1]$ but in fact one can describe the space $H^1(0,1)$ as the set of absolutely continuous functions u on the interval $[0,1]$ whose derivatives u' belong to $L_2(0,1)$.

Example 6.3. Let $n = 2$ or $n = 3$, $k = 0$ and $Q \subset \mathbb{R}^n$ be a bounded domain with C^2-boundary. Then the space $H^2(Q)$ is embedded in the space $C(Q)$, i. e., all functions from $H^2(Q)$ are continuous in Q. Such a conclusion cannot be made regarding functions from $H^1(Q)$. Indeed, the discontinuous function $u(x) = (x_1^2 + x_2^2 + x_3^2)^{\mu/2}$ with $-1/2 < \mu < 0$ from Example 5.3 belongs to the space $H^1(Q)$ in the unit ball $Q = B_1$ from $\mathbb{R}^3$. It is easy to give a counterexample confirming that functions from $H^1(Q)$ do not have to be continuous in the case of a two-dimensional domain Q. For any values of $n \geq 2$ and $k \geq 0$, one can easily verify that the integer exponent $l = k + [n/2] + 1$ is the smallest possible for embedding the corresponding Sobolev space into the space of k-times continuously-differentiable functions.

3 Elliptic problems

7 Dirichlet problem

Unique solvability

We begin this chapter with a surprisingly short proof of the unique solvability of the Dirichlet problem for the Poisson equation. But first we should say how a solution to the problem is understood. As before, Q denotes a bounded domain in $\mathbb{R}^n$ with C^1-boundary.

Definition 7.1. Classical solution to the problem

$$-\Delta u(x) = f(x) \quad (x \in Q), \tag{7.1}$$

$$u(x) = 0 \quad (x \in \partial Q) \tag{7.2}$$

is a function $u \in C^2(Q) \cap C(\overline{Q})$ that satisfies equation (7.1) at each point of the domain Q and condition (7.2) at each point of ∂Q, and the function f is assumed to be continuous in Q.

The fact that the boundary function in condition (7.2) is taken equal to zero is not a loss of generality if the function γ on the boundary can be extended into the domain by a function $u_\gamma \in C^2(Q) \cap C(\overline{Q})$ satisfying the Laplace equation $\Delta u_\gamma = 0$ in Q (a harmonic function). Then the difference $u - u_\gamma$ satisfies the Poisson equation and vanishes at the boundary.

Let us multiply (7.1) by an arbitrary function $\overline{v} \in C_0^\infty(Q)$ and integrate the resulting equality over Q. After integration by parts, we have

$$\int_Q \nabla u \nabla \overline{v}\, dx = \int_Q f \overline{v}\, dx. \tag{7.3}$$

If we assume that the function f lies in $L_2(Q)$, and our classical solution u belongs to the space $H^1(Q)$, then equation (7.3) can be extended to all functions $v \in \mathring{H}^1(Q)$. On the other hand, the requirements $f \in L_2(Q)$ and $u \in H^1(Q)$ alone are sufficient for the integrals in (7.3) to be well-defined. In addition, condition (7.2) is naturally interpreted in the sense of a trace on the boundary of a function from $H^1(Q)$.

The above considerations motivate the following important definition.

Definition 7.2. Let $f \in L_2(Q)$. A function $u \in \mathring{H}^1(Q)$ is called a *weak solution* of problem (7.1), (7.2) if integral identity (7.3) is satisfied for any function $v \in \mathring{H}^1(Q)$.

Note that by virtue of Corollary 6.1 the integral on the left-hand side of (7.3) specifies the equivalent inner product $(u, v)'_{\mathring{H}^1(Q)}$ in the space $\mathring{H}^1(Q)$.

https://doi.org/10.1515/9783112229637-003

Lemma 7.1. *The equality*

$$(f, v)_{L_2(Q)} = (Af, v)'_{\mathring{H}^1(Q)} \quad (f \in L_2(Q),\ v \in \mathring{H}^1(Q)) \tag{7.4}$$

defines a bounded linear operator $A : L_2(Q) \to \mathring{H}^1(Q)$. *The restriction of* A *to the space* $\mathring{H}^1(Q)$ *is a positive compact operator in* $\mathring{H}^1(Q)$.

Proof. The Cauchy–Schwarz and Friedrichs inequalities allow us to interpret the integral (f, v) as a continuous antilinear (conjugate-linear) functional with respect to the function v on the space $\mathring{H}^1(Q)$,

$$\left|(f, v)_{L_2(Q)}\right| \leqslant \|f\|_{L_2(Q)}\|v\|_{L_2(Q)} \leqslant d\|f\|_{L_2(Q)}\|v\|'_{\mathring{H}^1(Q)},$$

where $d = \operatorname{diam} Q$ is the constant from inequality (6.15). By Riesz's representation theorem (Theorem 2.7), there exists a unique function $F \in \mathring{H}^1(Q)$ such that $(f, v)_{L_2(Q)} = (F, v)'_{\mathring{H}^1(Q)}$ for all functions $v \in \mathring{H}^1(Q)$. In this case, we have $\|F\|'_{\mathring{H}^1(Q)} \leqslant d\|f\|_{L_2(Q)}$, i. e., the correspondence $f \mapsto F$ is a bounded linear operator from $L_2(Q)$ to $\mathring{H}^1(Q)$. We denote this operator by A.

The restriction of A to the space $\mathring{H}^1(Q)$ is a compact operator in $\mathring{H}^1(Q)$ due to the compact embedding of $\mathring{H}^1(Q)$ in $L_2(Q)$ (Theorem 6.5). Moreover, it follows from (7.4) that

$$(Au, v)'_{\mathring{H}^1(Q)} = (u, v)_{L_2(Q)} = \overline{(v, u)}_{L_2(Q)} = \overline{(Av, u)'}_{\mathring{H}^1(Q)} = (u, Av)'_{\mathring{H}^1(Q)}$$

for $u, v \in \mathring{H}^1(Q)$ and

$$(Av, v)'_{\mathring{H}^1(Q)} = \|v\|^2_{L_2(Q)} > 0, \quad v \neq 0.$$

In other words, the operator $A : \mathring{H}^1(Q) \to \mathring{H}^1(Q)$ is self-adjoint and positive. □

Now identity (7.3) takes the form

$$(u, v)'_{\mathring{H}^1(Q)} = (Af, v)'_{\mathring{H}^1(Q)} \quad (v \in \mathring{H}^1(Q)),$$

and it immediately follows that $u = Af$. We have proven the following theorem.

Theorem 7.1. *Problem* (7.1), (7.2) *has a unique weak solution* $u \in \mathring{H}^1(Q)$ *for any function* $f \in L_2(Q)$. *This solution satisfies the inequality*

$$\int_Q |\nabla u|^2 dx \leqslant (\operatorname{diam} Q)^2 \int_Q |f|^2 dx. \tag{7.5}$$

To conclude this section, we discuss the case of a nonhomogeneous boundary condition,

$$u|_{\partial Q} = \gamma(x) \quad (x \in \partial Q). \tag{7.6}$$

Definition 7.3. The function $u \in H^1(Q)$ is called a *weak solution* of problem (7.1), (7.6) if it satisfies identity (7.3) for any function $v \in \mathring{H}^1(Q)$ and boundary condition (7.6) in the sense of a trace, where $f \in L_2(Q)$.

It is clear that a necessary condition for the existence of a weak solution to problem (7.1), (7.6) is the possibility of extending the function γ into the domain Q by a function $u_\gamma \in H^1(Q)$. We call such function γ *admissible* (previously we used the term *trace space* in a similar situation, see Remark 6.3).

If the function γ is admissible, then by setting $w = u - u_\gamma$, the original problem is reduced to finding a function $w \in \mathring{H}^1(Q)$ satisfying the integral identity

$$\int_Q \nabla w \nabla \overline{v}\, dx = \int_Q (f\overline{v} - \nabla u_\gamma \nabla \overline{v})\, dx \tag{7.7}$$

for all $v \in \mathring{H}^1(Q)$. Now both functions v and w belong to the space $\mathring{H}^1(Q)$, where the integral from the left-hand side in (7.7) defines the equivalent inner product $(w, v)'_{\mathring{H}^1(Q)}$. Obvious inequalities

$$\left|\int_Q (f\overline{v} - \nabla u_\gamma \nabla \overline{v})\, dx\right| \leqslant \|f\|_{L_2(Q)}\|v\|_{L_2(Q)} + \||\nabla u_\gamma|\|_{L_2(Q)}\||\nabla v|\|_{L_2(Q)} \leqslant$$
$$\leqslant (d\|f\|_{L_2(Q)} + \|u_\gamma\|_{H^1(Q)})\|v\|'_{\mathring{H}^1(Q)}$$

show that the integral on the right-hand side in (7.7) is still a continuous antilinear functional in v in the space $\mathring{H}^1(Q)$. Therefore, the Riesz representation theorem guarantees a unique solution w of (7.7) with the estimate $\|w\|'_{\mathring{H}^1(Q)} \leqslant (d\|f\|_{L_2(Q)} + \|u_\gamma\|_{H^1(Q)})$. Using the obtained w, we pass to the function $u = w + u_\gamma$ being a weak solution to problem (7.1), (7.6). Moreover,

$$\|u\|_{H^1(Q)} \leqslant \|w\|_{H^1(Q)} + \|u_\gamma\|_{H^1(Q)} \leqslant \sqrt{d^2+1}\,\|w\|'_{\mathring{H}^1(Q)} + \|u_\gamma\|_{H^1(Q)} \leqslant$$
$$\leqslant d\sqrt{(d^2+1)}\|f\|_{L_2(Q)} + (1 + \sqrt{d^2+1})\|u_\gamma\|_{H^1(Q)}. \tag{7.8}$$

At first glance, it seems that the obtained weak solution depends on how the function γ is extended into the domain. In reality that is, of course, not the case. Let u and u' be weak solutions to problem (7.1), (7.6). Then the difference $\tilde{u} = u - u'$ belongs to $\mathring{H}^1(Q)$ and satisfies $(\tilde{u}, v)'_{\mathring{H}^1(Q)} = 0$ for all $v \in \mathring{H}^1(Q)$. Consequently, $\tilde{u} = 0$, i. e., $u = u'$.

One can pass in (7.8) to the infimum of all possible extensions u_γ of the function γ and obtain the following estimate for the weak solution u:

$$\|u\|_{H^1(Q)} \leqslant c_1(\|f\|_{L_2(Q)} + \|\gamma\|_{H^{1/2}(Q)}) \tag{7.9}$$

(see (6.7)).

Theorem 7.2. *Problem* (7.1), (7.6) *has a unique weak solution* $u \in H^1(Q)$ *for any* $f \in L_2(Q)$ *and admissible* γ *on* ∂Q. *This solution satisfies estimate* (7.9) *with a positive constant* c_1 *depending only on* Q.

Of course, this theorem is conditional until we have an explicit description of the trace space.

Eigenfunctions and eigenvalues

Along with (7.1), (7.2), we consider the corresponding problem on eigenfunctions and eigenvalues

$$-\Delta u(x) = \lambda u(x) \quad (x \in Q), \tag{7.10}$$

$$u|_{\partial Q} = 0. \tag{7.11}$$

A number $\lambda \in \mathbb{C}$ for which problem (7.10), (7.11) has a nontrivial solution u is called an *eigenvalue*, and the solution itself is called the *eigenfunction*. Let us give a precise definition in the spirit of the previous section.

Definition 7.4. A nonzero function $u \in \mathring{H}^1(Q)$ is called the *weak eigenfunction* of the Dirichlet problem for the Laplace operator, corresponding to the *eigenvalue* λ, if the integral identity

$$\int_Q \nabla u \nabla \overline{v}\, dx = \lambda \int_Q u \overline{v}\, dx \tag{7.12}$$

holds for all functions $v \in \mathring{H}^1(Q)$.

All eigenfunctions corresponding to the same eigenvalue form a linear space. Its dimension (in the general situation not necessarily finite) is called the *multiplicity* of the eigenvalue.

Theorem 7.3. *All eigenvalues of the Dirichlet problem for the operator* $-\Delta$ *are real, positive, isolated, have finite multiplicity and* $\lambda_s \to +\infty$ *as* $s \to \infty$. *There exists an orthonormal basis* u_s, $s = 1, 2, \ldots$, *in the space* $L_2(Q)$, *consisting of eigenfunctions of the Dirichlet problem. In addition, the functions* $u_s/\sqrt{\lambda_s}$, $s = 1, 2 \ldots$, *form an orthonormal basis in the space* $\mathring{H}^1(Q)$ *with the inner product*

$$(w_1, w_2)'_{\mathring{H}^1(Q)} = \int_Q \nabla w_1 \nabla \overline{w}_2\, dx.$$

Proof. Using the operator A from Lemma 7.1, we can rewrite identity (7.12) in the form of the operator equation $u = \lambda A u$ in the space $\mathring{H}^1(Q)$. Obviously, $\lambda = 0$ is not an eigenvalue of the problem. After dividing by λ, we arrive at the eigenvalue problem $Au = \mu u$ for the self-adjoint positive compact operator A in the Hilbert space $\mathring{H}^1(Q)$. Here, we denote $\mu = 1/\lambda$. By the Hilbert–Schmidt theorem (Theorem 2.16), Lemma 2.3 and Theorem 2.15, there exists a countable set of positive isolated eigenvalues μ_s, $s = 1, 2, \ldots$, of finite multiplicity, and $\mu_s \to 0$ as $s \to \infty$. We enumerate them in nongrowing order, taking into account their multiplicity, $\mu_1 \geqslant \mu_2 \geqslant \cdots \geqslant \mu_s \geqslant \cdots > 0$. The corresponding eigenvalues of the Dirichlet problem $\lambda_s = 1/\mu_s$ in this case form a nondecreasing sequence

$$0 < \lambda_1 \leqslant \lambda_2 \leqslant \cdots \leqslant \lambda_s \leqslant \ldots \to +\infty.$$

Each eigenvalue μ_s of the operator A can be associated with an eigenfunction u_s in a way that the system u_s, $s = 1, 2 \ldots$, forms an orthogonal basis in $\mathring{H}^1(Q)$ and $\|u_s\|_{L_2(Q)} = 1$. If u_{s_1} and u_{s_2} are eigenfunctions, then substituting u_{s_1} for u and u_{s_2} for v in formula (7.12), we have

$$(u_{s_1}, u_{s_2})'_{\mathring{H}^1(Q)} = \lambda_{s_1}(u_{s_1}, u_{s_2})_{L_2(Q)},$$

i. e., the system u_s is as well orthogonal in the space $L_2(Q)$ (take $s_1 \neq s_2$) and $\|u_s\|'_{\mathring{H}^1(Q)} = \sqrt{\lambda_s}$ (take $s_1 = s_2 = s$). From the density of $\mathring{H}^1(Q)$ in $L_2(Q)$, it follows that the system u_s is complete and, therefore, forms an orthonormal basis in $L_2(Q)$. Finally, note that the system $u_s/\sqrt{\lambda_s}$ is an orthonormal basis in $\mathring{H}^1(Q)$. □

Remark 7.1. The Hilbert–Schmidt theorem also provides a variational principle for the eigenfunctions and eigenvalues of the Laplacian with the Dirichlet condition. Thus, the largest eigenvalue μ_1 of a positive compact operator A is the maximum of the quadratic form $(Au, u)'_{\mathring{H}^1(Q)}$ on the unit sphere in $\mathring{H}^1(Q)$,

$$\mu_1 = \sup_{0 \neq u \in \mathring{H}^1(Q)} \frac{(Au, u)'_{\mathring{H}^1(Q)}}{\|u\|'^2_{\mathring{H}^1(Q)}}.$$

Accordingly, we obtain the following formula for the first (smallest) eigenvalue λ_1 of the Dirichlet problem:

$$\lambda_1 = \inf_{0 \neq u \in \mathring{H}^1(Q)} \frac{\|u\|'^2_{\mathring{H}^1(Q)}}{(Au, u)'_{\mathring{H}^1(Q)}} = \inf_{0 \neq u \in \mathring{H}^1(Q)} \frac{\|u\|'^2_{\mathring{H}^1(Q)}}{\|u\|^2_{L_2(Q)}} = \inf_{0 \neq u \in \mathring{H}^1(Q)} \frac{\||\nabla u|\|^2_{L_2(Q)}}{\|u\|^2_{L_2(Q)}},$$

and the minimum is reached when u coincides with the first eigenfunction u_1. A similar variational formulation is available for all other eigenvalues and eigenfunctions. A good exercise for the reader is to prove that the smallest eigenvalue of λ_1 is prime, i. e., has multiplicity 1, and the corresponding eigenfunction u_1 does not vanish in Q (the fact that all eigenfunctions are infinitely differentiable in the domain is shown later in this section).

Having the eigenfunctions $u_s(x)$ of the Dirichlet problem for the Laplace operator in Q, one can represent the solution to problem (7.1), (7.2) in the form of the Fourier series

$$u(x) = \sum_{s=1}^{\infty} c_s \frac{u_s(x)}{\sqrt{\lambda_s}}, \quad c_s = \left(u, \frac{u_s}{\sqrt{\lambda_s}}\right)'_{\mathring{H}^1(Q)} \quad (s = 1, 2, \ldots)$$

convergent in the space $\mathring{H}^1(Q)$. It is enough to set $v = u_s/\sqrt{\lambda_s}$ in equation (7.3) to obtain $c_s = (f, u_s/\sqrt{\lambda_s})_{L_2(Q)}$. Therefore,

$$u(x) = \sum_{s=1}^{\infty} \frac{(f, u_s)_{L_2(Q)}}{\lambda_s} u_s(x).$$

Fredholm solvability

The Dirichlet problem for a more general elliptic equation

$$-\Delta u(x) + \sum_{i=1}^{n} a_i(x) u_{x_i}(x) + a_0(x) u(x) = f(x), \tag{7.13}$$

$$u|_{\partial Q} = 0 \tag{7.14}$$

may no longer be uniquely solvable. But the situation here is still “not too bad” and is described by *Fredholm’s theorems.*

Along with problem (7.13), (7.14), we consider the corresponding homogeneous problem

$$-\Delta v(x) + \sum_{i=1}^{n} a_i(x) v_{x_i}(x) + a_0(x) v(x) = 0, \tag{7.15}$$

$$v|_{\partial Q} = 0 \tag{7.16}$$

and the homogeneous adjoint (formally for now) problem

$$-\Delta w(x) - \left(\sum_{i=1}^{n} a_i(x) w(x)\right)_{x_i} + a_0(x) w(x) = 0, \tag{7.17}$$

$$w|_{\partial Q} = 0. \tag{7.18}$$

We assume that $a_i \in C^1(\overline{Q})$, $i = 1, \ldots, n$, $a_0 \in C(\overline{Q})$ and $f \in L_2(Q)$. We consider the coefficients to be real-valued functions, and the function f to be complex valued.

As in the case of the Poisson equation with the Dirichlet condition, we can define the weak solution to problem (7.13), (7.14) using an integral identity.

Definition 7.5. A function $u \in \mathring{H}^1(Q)$ is called a *weak solution* of problem (7.13), (7.14) if the integral identity

$$\int_Q \left(\nabla u \nabla \overline{\varphi} + \sum_{i=1}^n a_i u_{x_i} \overline{\varphi} + a_0 u \overline{\varphi} \right) dx = \int_Q f \overline{\varphi}\, dx \tag{7.19}$$

holds for all functions $\varphi \in \mathring{H}^1(Q)$. Similarly, the identities

$$\int_Q \left(\nabla v \nabla \overline{\varphi} + \sum_{i=1}^n a_i v_{x_i} \overline{\varphi} + a_0 v \overline{\varphi} \right) dx = 0$$

and

$$\int_Q \left(\nabla w \nabla \overline{\varphi} - \sum_{i=1}^n (a_i w)_{x_i} \overline{\varphi} + a_0 w \overline{\varphi} \right) dx = 0$$

on the space $\mathring{H}^1(Q)$ define weak solutions $v \in \mathring{H}^1(Q)$ and $w \in \mathring{H}^1(Q)$ of homogeneous problems (7.15), (7.16) and (7.17), (7.18), respectively.

Theorem 7.4. (a) *If homogeneous problem* (7.15), (7.16) *has the only trivial solution, then nonhomogeneous problem* (7.13), (7.14) *has a unique weak solution for any function* $f \in L_2(Q)$.

(b) *Suppose that homogeneous problem* (7.15), (7.16) *has a nontrivial solution. In this case, nonhomogeneous problem* (7.13), (7.14) *is solvable if and only if the function* f *is orthogonal in* $L_2(Q)$ *to each weak solution to homogeneous adjoint problem* (7.17), (7.18).

(c) *Homogeneous problems* (7.15), (7.16) *and* (7.17), (7.18) *have the same finite number of linearly independent weak solutions.*

Proof. Consider the sesquilinear form

$$\Phi(u, \varphi) = \int_Q \left(\sum_{i=1}^n a_i u_{x_i} + a_0 u \right) \overline{\varphi}\, dx = \int_Q \left(-u \sum_{i=1}^n (a_i \overline{\varphi})_{x_i} + a_0 u \overline{\varphi} \right) dx$$

on the space $\mathring{H}^1(Q)$. The conditions imposed on the coefficients $a_0, a_1, \dots, a_n$ and the Friedrichs inequality imply the estimate $|\Phi(u, \varphi)| \leqslant k_1 \|u\|_{L_2(Q)} \|\varphi\|'_{\mathring{H}^1(Q)}$. By the Riesz representation theorem, there exists a bounded linear operator $T : L_2(Q) \to \mathring{H}^1(Q)$ such that $\Phi(u, \varphi) = (Tu, \varphi)'_{\mathring{H}^1(Q)}$ and $\|Tu\|'_{\mathring{H}^1(Q)} \leqslant k_1 \|u\|_{L_2(Q)}$. As an operator from $\mathring{H}^1(Q)$ to $\mathring{H}^1(Q)$, T is compact (see Theorem 6.5).

In view of Lemma 7.1, integral identity (7.19) takes the form

$$((I + T)u, \varphi)'_{\mathring{H}^1(Q)} = (Af, \varphi)'_{\mathring{H}^1(Q)} \quad (\varphi \in \mathring{H}^1(Q)).$$

Therefore, a function $u \in \mathring{H}^1(Q)$ is a weak solution to problem (7.13), (7.14) if and only if it is a solution to the operator equation

$$(I + T)u = Af \tag{7.20}$$

with the compact operator T in the Hilbert space $\mathring{H}^1(Q)$. Corresponding homogeneous problem (7.15), (7.16) has the following form:

$$(I + T)v = 0. \tag{7.21}$$

As for homogeneous adjoint problem (7.17), (7.18), integrating by parts we obtain

$$\int_Q \left(-\sum_{i=1}^n (a_i w)_{x_i} + a_0 w \right) \overline{\varphi}\, dx = \int_Q \left(\sum_{i=1}^n a_i \overline{\varphi}_{x_i} + a_0 \overline{\varphi} \right) w\, dx =$$
$$= \overline{(T\varphi, w)'_{\mathring{H}^1(Q)}} = \overline{(\varphi, T^* w)'_{\mathring{H}^1(Q)}} = (T^* w, \varphi)'_{\mathring{H}^1(Q)},$$

where T^* denotes the adjoint operator with respect to the equivalent inner product in $\mathring{H}^1(Q)$; see Corollary 6.1. Therefore, problem (7.17), (7.18) is equivalent to the adjoint operator equation

$$(I + T^*)w = 0. \tag{7.22}$$

The rest of the proof is the content of the well-known Fredholm theorems (see Section 2).

(a) Suppose that homogeneous problem (7.15), (7.16) has the only trivial solution, i. e., the only solution to homogeneous equation (7.21) is the trivial one, $v = 0$. But then nonhomogeneous equation (7.20) is uniquely solvable for any function $Af \in \mathring{H}^1(Q)$, i. e., problem (7.13), (7.14) has a unique weak solution for any function $f \in L_2(Q)$.

(b) Let there exist nontrivial solutions to homogeneous problem (7.15) and (7.16), i. e., homogeneous equation (7.21) has a nontrivial solution. In this case, nonhomogeneous equation (7.20) has a solution if and only if $(Af, w)'_{\mathring{H}^1(Q)} = 0$ for all solutions w of homogeneous adjoint equation (7.22). Since $(Af, w)'_{\mathring{H}^1(Q)} = (f, w)_{L_2(Q)}$ (Lemma 7.1), problem (7.13), (7.14) has a weak solution if and only if $(f, w)_{L_2(Q)} = 0$ for all weak solutions w of homogeneous adjoint problem (7.17), (7.18).

(c) Obviously, the solution spaces of homogeneous problems (7.15), (7.16) and (7.17), (7.18) are exactly the kernels of the linear operators $I+T$ and $I+T^*$, dimensions of which are finite and identical. □

A typical example of an application of Theorem 7.4 is the following statement.

Corollary 7.1. *Let the coefficients of equation* (7.13) *satisfy the following inequality in the domain Q:*

$$a_0(x) \geqslant \frac{1}{2}\sum_{i=1}^{n} a_{ix_i}(x). \tag{7.23}$$

Then problem (7.13), (7.14) *has a unique weak solution for any function* $f \in L_2(Q)$.

Proof. Make sure that the homogeneous problem has the only trivial solution. Then the statement to be proved follows from part (a) of Theorem 7.4.

It follows from the equality $v + Tv = 0$ that

$$0 = (v + Tv, v)'_{\mathring{H}^1(Q)} = \|v\|'^2_{\mathring{H}^1(Q)} + \Phi(v, v)$$

and, therefore, $\|v\|'^2_{\mathring{H}^1(Q)} + \operatorname{Re}\Phi(v, v) = 0$. It suffices to prove the inequality

$$\operatorname{Re}\Phi(v, v) = \operatorname{Re}\int\limits_Q \left(\sum_{i=1}^{n} a_i v_{x_i} + a_0 v\right)\overline{v}\,dx \geqslant 0.$$

Integrating by parts and taking into account that $v|_{\partial Q} = 0$, we obtain

$$\int\limits_Q a_i v_{x_i}\overline{v}\,dx = -\int\limits_Q v(a_i\overline{v})_{x_i}\,dx = -\int\limits_Q a_i\overline{v}_{x_i} v\,dx - \int\limits_Q a_{ix_i}|v|^2\,dx.$$

Therefore,

$$2\operatorname{Re}\int\limits_Q a_i v_{x_i}\overline{v}\,dx = -\int\limits_Q a_{ix_i}|v|^2\,dx$$

and

$$\operatorname{Re}\Phi(v, v) = \int\limits_Q \left(a_0 - \frac{1}{2}\sum_{i=1}^{n} a_{ix_i}\right)|v|^2\,dx \geqslant 0$$

due to inequality (7.23). □

Remark 7.2. All results of this section remain valid if in equation (7.13) we replace the Laplacian with a more general elliptic operator

$$-\sum_{i,j=1}^{n} (a_{ij}(x)u_{x_j}(x))_{x_i},$$

where the coefficients $a_{ij}(x) = a_{ji}(x)$ are real-valued continuous functions in $\overline{Q}$ and the matrix composed of these coefficients is uniformly positive-definite with respect to $x \in \overline{Q}$, i. e., there exists a positive constant κ such that

$$\sum_{i,j=1}^{n} a_{ij}(x)\eta_i\overline{\eta}_j \geqslant \kappa|\eta|^2 \tag{7.24}$$

for all $\eta \in \mathbb{C}^n$ and $x \in \overline{Q}$, where $\kappa > 0$ does not depend on η and x.

The weak solution of the Dirichlet problem for the equation

$$-\sum_{i,j=1}^{n}(a_{ij}(x)u_{x_j}(x))_{x_i} + \sum_{i=1}^{n} a_i(x)u_{x_i}(x) + a_0(x)u(x) = f(x) \quad (x \in Q), \tag{7.25}$$

$f \in L_2(Q)$, is introduced in a similar way as a function $u \in \mathring{H}^1(Q)$ satisfying the identity

$$\int_Q \left(\sum_{i,j=1}^{n} a_{ij}u_{x_j}\overline{v}_{x_i} + \sum_{i=1}^{n} a_i u_{x_i}\overline{v} + a_0 u\overline{v} \right) dx = \int_Q f\overline{v}\, dx \quad (v \in \mathring{H}^1(Q)), \tag{7.26}$$

where $a_{ij} = a_{ji}$, $a_i \in C^1(\overline{Q})$ and $a_0 \in C(\overline{Q})$ are real-valued functions.

The key point in the reasoning is that the sesquilinear form

$$(u, v)'_{\mathring{H}^1(Q)} = \int_Q \sum_{i,j=1}^{n} a_{ij}u_{x_j}\overline{v}_{x_i}\, dx \quad (u, v \in \mathring{H}^1(Q)) \tag{7.27}$$

still defines an inner product on the space $\mathring{H}^1(Q)$ equivalent to the standard one. This is directly evident from algebraic inequality (7.24).

Variational formulation—Ritz method

Below we consider the variational problem for an abstract quadratic functional and establish a connection between solutions of the variational problem and weak solutions of the boundary-value problem for an elliptic equation. Since many laws of physics, mechanics and other sciences are formulated on the basis of variational principles, it becomes clear why using a weak solution is sometimes more natural than a classical one. We also present an approximate method for solving variational problems called the Ritz method. This method is also suitable for the approximate solution of elliptic boundary-value problems.

Let us first present the variational problem in an abstract situation. Let H be a separable real Hilbert space and l be a continuous linear functional on H, $l \in H^*$. Consider a quadratic functional E given by the formula

$$E(u) = \|u\|^2 - 2l(u), \quad u \in H.$$

It is often called the *energy functional*. Obvious inequality

$$E(u) \geqslant \|u\|^2 - 2\|l\|\|u\| \geqslant -\|l\|^2$$

means that the functional $E(u)$ is bounded from below. Here, $\|u\|$ denotes the norm in H, and $\|l\|$ is the norm of the functional in H^*. Let $d = \inf_{u\in H} E(u)$.

Definition 7.6. The sequence $u_s \in H$, $s = 1, 2\ldots$, is called *minimizing* for the functional E if $E(u_s) \to d$ as $s \to \infty$. An element $u_0 \in H$ is called *minimizing* if $E(u_0) = d$.

Obviously, the functional E always has a minimizing sequence. However, the existence of a minimizing element is not obvious.

Theorem 7.5. *There exists a unique element $u_0 \in H$ that minimizes the functional E. Any sequence in H that minimizes the functional E converges to u_0 in H.*

Proof. Let us show that every minimizing sequence u_s, $s = 1, 2\ldots$, is a Cauchy sequence in H.

Having $\varepsilon > 0$ fixed, choose a number $N = N(\varepsilon)$ such that $d \leqslant E(u_s) \leqslant d + \varepsilon$ for $s \geqslant N$. Using the parallelogram identity and then the definition of $E(u)$, we write

$$\left\| \frac{u_m - u_s}{2} \right\|^2 = \frac{\|u_m\|^2 + \|u_s\|^2}{2} - \left\| \frac{u_m + u_s}{2} \right\|^2 = \\ = \frac{E(u_m) + E(u_s)}{2} - E\left(\frac{u_m + u_s}{2} \right) \leqslant \frac{1}{2}(d + \varepsilon + d + \varepsilon) - d = \varepsilon,$$

if $s \geqslant N$ and $m \geqslant N$ due to the inequality $E((u_s + u_m)/2) \geqslant d$. Therefore, the sequence u_s is a Cauchy sequence.

Since the space H is complete, there exists an element $u_0 \in H$ such that $u_s \to u_0$ in H as $s \to \infty$. This immediately implies the relations $\|u_s\| \to \|u_0\|$ and $l(u_s) \to l(u_0)$, providing $E(u_s) \to E(u_0)$ as $s \to \infty$. On the other hand, $E(u_s) \to d$. Due to the uniqueness of the limit, we have $E(u_0) = d$. The existence of a minimizing element has been proven. If u_0 and u_0' are different minimizing elements, then the sequence $u_0, u_0', u_0, u_0', \ldots$ are both minimizing and divergent, which is impossible. □

Therefore, an approximate solution to the variational problem

$$E(u) \to \min, \quad u \in H,$$

can be obtained with the use of a minimizing sequence.

Let us now present the *Ritz method* for constructing a minimizing sequence.

Let a system $\varphi_1, \varphi_2, \ldots$ be linearly-independent and complete in H. Denoting the linear span of elements $\varphi_1, \varphi_2, \ldots, \varphi_s$ by R_s, we construct an expanding chain $R_1 \subset R_2 \subset \ldots$ of finite-dimensional subspaces, $\dim R_s = s$. Let us define u_s as the element that minimizes $E(u)$ on R_s. In order to find u_s, we seek a minimum of the quadratic function

$$F(c_1, \ldots, c_s) = E(c_1\varphi_1 + \cdots + c_s\varphi_s) = \sum_{i,j=1}^{s} (\varphi_i, \varphi_j)c_i c_j - 2\sum_{i=1}^{s} l(\varphi_i)c_i$$

depending on s real variables $c_1, \ldots, c_s$. The necessary minimum condition (in the case under consideration it is, of course, sufficient as well) $\partial F/\partial c_i = 0$ $(i = 1, \ldots, s)$ leads to the system

$$\sum_{j=1}^{s}(\varphi_i,\varphi_j)c_j = l(\varphi_i), \quad (i = 1,\dots,s) \tag{7.28}$$

of linear equations. The determinant of this system, being a Gram determinant, is nonzero, thus (7.28) always has a unique solution $c_1^s,\dots,c_s^s$. Then the function $u_s = c_1^s\varphi_1 + \dots + c_s^s\varphi_s$ provides a minimum to the functional $E(u)$ on the finite-dimensional space R_s. The constructed sequence u_s is called the *Ritz sequence* for $E(u)$ with respect to the system $\varphi_1,\varphi_2,\dots$. Obviously, $E(u_1) \geqslant E(u_2) \geqslant \dots$.

Theorem 7.6. *The Ritz sequence is minimizing.*

Proof. Let u minimize the functional $E(u)$. Choose an arbitrary $\varepsilon > 0$. Since the system $\varphi_1,\varphi_2,\dots$ is complete in H, there exists a number s and an element $u_\varepsilon \in R_s$ such that $\|u-u_\varepsilon\| < \varepsilon$. Since u_s minimizes $E(u)$ on R_s, we have

$$\begin{aligned} d \leqslant E(u_s) \leqslant E(u_\varepsilon) &= \|u_\varepsilon\|^2 - 2l(u_\varepsilon) = \|u_\varepsilon - u + u\|^2 - 2l(u_\varepsilon - u + u) = \\ &= E(u) + \|u_\varepsilon - u\|^2 - 2l(u_\varepsilon - u) + 2(u_\varepsilon - u, u) \leqslant d + \varepsilon^2 + 2(\|l\| + \|u\|)\varepsilon. \end{aligned}$$ □

Below we study a connection between the variational problem for the functional

$$E(u) = \int_Q (|\nabla u|^2 - 2fu)\,dx \quad (u \in \mathring{H}^1(Q)) \tag{7.29}$$

and boundary-value problem (7.1), (7.2).

Taking an arbitrary $v \in H$, consider the quadratic function

$$P(t) = E(u+tv) = E(u) + 2((u,v) - l(v))t + \|v\|^2t^2$$

of the real variable t. If the minimum of E is obtained on u, then the function $P(t)$ reaches its minimum value at the point $t = 0$, i. e., $(u,v) - l(v) = 0$. Conversely, if an element $u \in H$ is such that the equality

$$(u,v) = l(v), \quad v \in H, \tag{7.30}$$

holds for all elements $v \in H$, then $E(u+v) = E(u) + \|v\|^2 > E(u)$ for $v \neq 0$. But then E has the strict global minimum at u. Therefore, relation (7.30) serves as an equivalent definition of the minimizing element u.

Note that relation (7.30) turns into identity (7.3), which defines a weak solution to the Dirichlet problem for the Poisson equation if $H = \mathring{H}^1(Q)$ with $(u,v) = \int_Q \nabla u \nabla v\,dx$ and $l(v) = \int_Q fv\,dx$, where $f \in L_2(Q)$. We obtained the following statement.

Theorem 7.7. *There exists a unique function $u \in \mathring{H}^1(Q)$ that minimizes functional* (7.29) *on the space $\mathring{H}^1(Q)$. A function $u \in \mathring{H}^1(Q)$ gives a minimum to functional* (7.29) *if and only if it is a weak solution of problem* (7.1), (7.2).

Now we conclude that any Ritz sequence constructed for functional (7.29) converges in $\mathring{H}^1(Q)$ to the weak solution of problem (7.1), (7.2). The Ritz method thereby makes it possible to obtain an approximate solution to the elliptic boundary-value problem.

Smoothness of weak solutions

Let $Q \subset \mathbb{R}^n$ be a bounded domain (without any assumptions about the boundary for now) and $f \in L_2(Q)$. Let $H^k_{\mathrm{loc}}(Q)$ denote the linear space consisting of all functions f in Q such that $f \in H^k(\Omega)$ for any strictly interior subdomain Ω, $\overline{\Omega} \subset Q$. We begin this paragraph with a statement about the internal smoothness of weak solutions of elliptic problems. Let us consider equation (7.25), as well as associated integral identity (7.26). We assume that the functions $a_{ij} = a_{ji} \in C^1(\overline{Q})$ and $a_i, a_0 \in C(\overline{Q})$ are real-valued, and inequality (7.24) is satisfied for all $\eta \in \mathbb{C}^n$ and $x \in \overline{Q}$.

Theorem 7.8. *Let $u \in H^1(Q)$ satisfy integral identity* (7.26) *for any function $v \in \mathring{H}^1(Q)$, where $f \in L_2(Q)$. Then the function u belongs to $H^2_{\mathrm{loc}}(Q)$ and satisfies equation* (7.25) *a. e. in Q.*

Proof. Since $u \in H^1(Q)$, the function $\sum_i a_i u_{x_i} + a_0 u$ belongs to $L_2(Q)$. Then u satisfies the integral identity

$$\int_Q \sum_{i,j} a_{ij} u_{x_j} \overline{v}_{x_i}\, dx = \int_Q F \overline{v}\, dx, \tag{7.31}$$

where $F = f - \sum_i a_i u_{x_i} - a_0 u \in L_2(Q)$.

For an arbitrary domain Ω, $\overline{\Omega} \subset Q$, we introduce a cut-off function $\xi \in C_0^\infty(Q)$ such that $0 \leqslant \xi(x) \leqslant 1$, $\xi(x) = 1$ for $x \in \Omega$ and $\operatorname{supp} \xi \subset Q_{3d}$, where $d = \rho(\Omega, \partial Q)/4$. Consider the function $v = \xi v_0$, where $v_0 \in H^1(Q)$. Substituting this function for v in equation (7.31) and using the Leibniz formula twice, we obtain

$$\begin{aligned}\int_Q \sum_{i,j} a_{ij} (\xi u)_{x_j} \overline{v}_{0x_i}\, dx = \int_Q F\xi \overline{v}_0\, dx - \int_Q \sum_{i,j} a_{ij} u_{x_j} \xi_{x_i} \overline{v}_0\, dx\, + \\ + \int_Q \sum_{i,j} a_{ij} u \xi_{x_j} \overline{v}_{0x_i}\, dx.\end{aligned} \tag{7.32}$$

Then we set $v_0(x) = \delta^l_{-h} v_1(x)$, where $1 \leqslant l \leqslant n$, $0 < |h| < d$, $v_1 \in H^1(Q)$ and $\rho(\operatorname{supp} v_1, \partial Q) > 2d$. Here, δ^l_{-h} denotes the finite-difference operator introduced on page 45.

Let functions $a \in C^1(\overline{Q})$ and $w \in H^1(Q)$ be such that $\rho(\operatorname{supp} w, \partial Q) > 2d$, and $0 < |h| < d$. We take advantage of the obvious equality

$$\delta^l_h(aw) = a\delta^l_h w + w^l_h \delta^l_h a, \tag{7.33}$$

where $w_h^l = w(x_1, \dots, x_{l-1}, x_l + h, x_{l+1}, \dots, x_n)$.

Substituting $v_0 = \delta_{-h}^l v_1$ in (7.32) and using formulas (5.8) and (7.33), we arrive at the identity

$$\begin{aligned}
&\int_Q \sum_{i,j} a_{ij} (\delta_h^l(\xi u))_{x_j} \overline{v}_{1x_i}\, dx = \\
&= \int_Q \sum_{i,j} a_{ij} u_{x_j} \xi_{x_i} \delta_{-h}^l \overline{v}_1\, dx + \int_Q \sum_{i,j} \delta_h^l (u a_{ij} \xi_{x_j}) \overline{v}_{1x_i}\, dx - \\
&- \int_Q \sum_{i,j} (\delta_h^l a_{ij})((\xi u)_{x_j})_h^l \overline{v}_{1x_i}\, dx - \int_Q F\xi \delta_{-h}^l \overline{v}_1\, dx.
\end{aligned} \tag{7.34}$$

Let us denote the integral on the left-hand side of (7.34) by I_0, and the corresponding integrals on the right-hand side by I_1, I_2, I_3 and I_4. Using the Cauchy–Schwarz inequality and Theorem 5.5, we obtain

$$|I_4| \leqslant \|F\|_{L_2(Q)} \||\nabla v_1|\|_{L_2(Q)}, \tag{7.35}$$

$$|I_s| \leqslant k_s \|u\|_{H^1(Q)} \||\nabla v_1|\|_{L_2(Q)} \quad (s = 1, 2, 3), \tag{7.36}$$

where $k_s > 0$ $(s = 1, 2, 3)$ do not depend on v_1 and u.

Finally, set $v_1 = \delta_h^l(\xi u)$. Then

$$I_0 \geqslant \kappa \int_Q |\nabla(\delta_h^l(\xi u))|^2\, dx = \kappa \||\nabla(\delta_h^l(\xi u))|\|^2_{L_2(Q)} \tag{7.37}$$

due to ellipticity condition (7.24). It follows from (7.35)–(7.37) that

$$\kappa \||\nabla(\delta_h^l(\xi u))|\|_{L_2(Q)} \leqslant k_4 (\|F\|_{L_2(Q)} + \|u\|_{H^1(Q)}), \tag{7.38}$$

where $k_4 > 0$ does not depend on h, F and u.

Inequality (7.38) and Theorem 5.5 imply the existence of weak derivatives $(\xi u)_{x_p x_q} \in L_2(Q)$ $(p, q = 1, \dots, n)$. Consequently, there exist weak derivatives $u_{x_p x_q} \in L_2(\Omega)$, and

$$\|u_{x_p x_q}\|_{L_2(\Omega)} \leqslant k(\|f\|_{L_2(Q)} + \|u\|_{H^1(Q)}). \tag{7.39}$$

Therefore, $u \in H^2(\Omega)$.

It remains to prove that the function $u(x)$ satisfies equation (7.25) a. e. in Q. In order to do this, take $v \in C_0^\infty(\Omega)$ in (7.26) for an arbitrary domain $\Omega \subset \mathbb{R}^n$, $\overline{\Omega} \subset Q$. Then, based on the definition of the weak derivative, we have

$$\int_\Omega \Big(-\sum_{i,j} (a_{ij} u_{x_j})_{x_i} + \sum_i a_i u_{x_i} + a_0 u - f \Big) \overline{v}\, dx = 0,$$

i. e., equation (7.25) is valid a. e. in Q. □

Remark 7.3. The result obtained can be enhanced.

Let $a_{ij}, a_i, a_0 \in C^\infty(\overline{Q})$. Suppose that $f \in L_2(Q) \cap H^k_{\mathrm{loc}}(Q)$ and $u \in H^1(Q)$ satisfies integral identity (7.26) *for all $v \in \mathring{H}^1(Q)$. Then $u \in H^{2+k}_{\mathrm{loc}}(Q)$.*

Combining Theorem 7.1, Theorem 7.8 and inequality (7.39), we obtain the following statement.

Corollary 7.2. *Let $f \in L_2(Q)$. Then a weak solution u of problem* (7.1) *and* (7.2) *belongs to the space $H^2_{\mathrm{loc}}(Q)$, and for any domain $\Omega \subset \mathbb{R}^n$ such that $\overline{\Omega} \subset Q$, the estimate*

$$\|u\|_{H^2(\Omega)} \leqslant c_0 \|f\|_{L_2(Q)} \tag{7.40}$$

holds, where c_0 does not depend on f.

It is natural to expect that the smoothness of a weak solution near the boundary depends on the smoothness of the boundary itself (as well as on the smoothness of a given function in the boundary condition if the latter is nonhomogeneous).

Theorem 7.9. *Let $Q \subset \mathbb{R}^n$ be a bounded domain, $\partial Q \in C^2$ and $f \in L_2(Q)$. Then a weak solution $u \in \mathring{H}^1(Q)$ of problem* (7.25), (7.2) *belongs to the space $H^2(Q)$.*

Proof. Taking Theorems 5.4 and 7.8 into account, we only need to prove that for any point $x^0 \in \partial Q$ there exists $a > 0$ such that $u \in H^2(Q \cap B_a(x^0))$.

Since $\partial Q \in C^2$, there exists a number $\delta > 0$ and a C^2-diffeomorphism ω of the ball $B_\delta(x^0)$ onto a neighborhood V of the origin, with the Jacobian equal to 1, such that

$$\omega(B_\delta(x^0) \cap Q) = \Omega = \{y \in V : y_n > 0\},$$
$$\omega(B_\delta(x^0) \cap \partial Q) = \{y \in V : y_n = 0\},$$

and $\omega(x_0) = 0$ (see Section 1).

Let us introduce a cut-off function $\xi \in C^\infty(\mathbb{R}^n)$, $\xi(x) = 1$ for $x \in B_{\delta/4}(x^0)$, $\operatorname{supp} \xi \subset B_{\delta/2}(x^0)$. In integral identity (7.31), we set $v = \xi v_0$, where v_0 is now an arbitrary function from $H^1(B_\delta(x^0) \cap Q)$ such that $v_0|_{\partial Q \cap B_\delta(x^0)} = 0$. Applying the Leibniz formula, we write

$$\int_Q \sum_{i,j} a_{ij}(\xi u)_{x_j} \overline{v}_{0x_i}\, dx = \int_Q F\xi \overline{v}_0\, dx - \int_Q \sum_{i,j} a_{ij} u_{x_j} \xi_{x_i} \overline{v}_0\, dx + \\ + \int_Q \sum_{i,j} a_{ij} u \xi_{x_j} \overline{v}_{0x_i}\, dx \tag{7.41}$$

similar to (7.32), where $F = f - \sum a_i u_{x_i} - a_0 u \in L_2(Q)$.

Let us pass to the local coordinates y in (7.41). According to the chain rule,

$$\frac{\partial}{\partial x_i} = \sum_p \frac{\partial y_p}{\partial x_i} \frac{\partial}{\partial y_p},$$

we have

$$\int_\Omega \sum_{p,q} b_{pq}(\xi u)_{y_q}\overline{v}_{0y_p}\,dy = \int_\Omega F\xi\overline{v}_0\,dy - \int_\Omega \sum_{p,q} b_{pq}u_{y_q}\xi_{y_p}\overline{v}_0\,dy + \\ + \int_\Omega u \sum_{p,q} b_{pq}\xi_{y_q}\overline{v}_{0y_p}\,dy \tag{7.42}$$

(we have preserved the same notation for functions). Here,

$$b_{pq} = \sum_{i,j} \frac{\partial y_p}{\partial x_i}\, a_{ij}\, \frac{\partial y_q}{\partial x_j} \in C^1(\Omega).$$

By definition, the matrix $B(y) = (b_{pq}(y))_{p,q=1}^n$ is symmetric and is such that

$$\sum_{p,q} b_{pq}(y)\eta_p\overline{\eta}_q \geqslant k_0|\eta|^2 \quad (y \in \Omega,\ \eta \in \mathbb{C}^n), \tag{7.43}$$

where $k_0 > 0$ does not depend on η and y.

Denote $v_0(y) = \delta^l_{-h}v_1(y)$, where $1 \leqslant l \leqslant n-1$,

$$0 < |h| < d = \rho(\omega(B_{\delta/2}(x^0)), \partial V)/2,$$

$v_1 \in H^1(\Omega)$, $v_1|_{\partial\Omega\cap\{y_n=0\}} = 0$ and $\rho(\operatorname{supp} v_1, \partial V) > d$. For arbitrary functions $b \in C^1(\overline{\Omega}\setminus\partial V)$ and $w \in H^1(\Omega)$ such that $\rho(\operatorname{supp} w, \partial V) \geqslant 2d$ and $0 < |h| < d$, the equality

$$\delta^l_h(bw) = b\delta^l_h w + w^l_h\delta^l_h b \tag{7.44}$$

is obvious, where $w^l_h = w(y_1, \dots, y_{l-1}, y_l + h, y_{l+1}, \dots, y_n)$.

Substituting $\delta^l_{-h}v_1$ for v_0 in relation (7.42) and using (7.44) together with (5.8), we obtain the following identity:

$$\int_\Omega \sum_{p,q} b_{pq}(\delta^l_h(\xi u))_{y_q}\overline{v}_{1y_p}\,dy = \int_\Omega \sum_{p,q} b_{pq}u_{y_q}\xi_{y_p}\delta^l_{-h}\overline{v}_1\,dy + \\ + \int_\Omega \sum_{p,q} \delta^l_h(ub_{pq}\xi_{y_q})\overline{v}_{1y_p}\,dy - \int_\Omega \sum_{p,q} (\delta^l_h b_{pq})((\xi u)_{y_q})^l_h\overline{v}_{1y_p}\,dy - \\ - \int_\Omega F\xi\delta^l_{-h}\overline{v}_1\,dy. \tag{7.45}$$

Let I_0 denote the integral on the left-hand side of (7.45), and I_1, I_2, I_3 and I_4 the corresponding integrals on the right-hand side. Applying the Cauchy–Schwarz inequality and Theorem 5.6, we see that

$$|I_j| \leqslant k_j\|u\|_{H^1(\Omega)}\||\nabla v_1|\|_{L_2(\Omega)} \quad (j = 1, 2, 3), \tag{7.46}$$

$$|I_4| \leqslant \|F\|_{L_2(\Omega)} \| |\nabla v_1| \|_{L_2(\Omega)}, \tag{7.47}$$

where $k_j > 0$ does not depend on v_1 and u.

Now set $v_1 = \delta_h^l(\xi u)$. Then inequality (7.43) provides the estimate

$$I_0 \geqslant k_0 \int\limits_{\Omega} |\nabla \delta_h^l(\xi u)|^2 \, dy = k_0 \| |\nabla \delta_h^l(\xi u)| \|^2_{L_2(\Omega)}, \tag{7.48}$$

where $k_0 > 0$ does not depend on h and u.

Combining relations (7.46)–(7.48), we arrive at the inequality

$$\| |\nabla \delta_h^l(\xi u)| \|_{L_2(\Omega)} \leqslant k_4 (\|f\|_{L_2(\Omega)} + \|u\|_{H^1(\Omega)}), \tag{7.49}$$

where $k_4 > 0$ does not depend on h, f and u.

Together with Theorem 5.6, inequality (7.49) implies the existence of weak derivatives $(\xi u)_{y_p y_q} \in L_2(\Omega)$ and

$$\|(\xi u)_{y_p y_q}\|_{L_2(\Omega)} \leqslant k_4 (\|f\|_{L_2(\Omega)} + \|u\|_{H^1(\Omega)}), \tag{7.50}$$

where $p+q < 2n$. This proves the existence of weak derivatives $u_{y_p y_q} \in L_2(\Omega_0)$ $(p+q < 2n)$ along with the estimate

$$\|u_{y_p y_q}\|_{L_2(\Omega_0)} \leqslant k_4 (\|f\|_{L_2(\Omega)} + \|u\|_{H^1(\Omega)}), \tag{7.51}$$

where $\Omega_0 = \{y \in \omega(B_{\delta/4}(x^0)) : y_n > 0\}$.

Let us now consider (7.42) for an arbitrary function $v_0 \in C_0^\infty(\Omega_0)$. Note that the function $\xi(x(y))$ is identically 1 on $\operatorname{supp} v_0$, thus we can write

$$\int\limits_{\Omega_0} \sum_{p,q} b_{pq} u_{y_q} \overline{v}_{0y_p} \, dy = \int\limits_{\Omega_0} F \overline{v}_0 \, dy. \tag{7.52}$$

By Theorem 7.8, the function u belongs to $H^2_{\mathrm{loc}}(\Omega_0)$ and satisfies the equation

$$-\sum_{p,q} (b_{pq}(y) u_{y_q}(y))_{y_p} = F(y) \tag{7.53}$$

a. e. in Ω_0. Since the matrix $B(y)$ is positive-definite in $\overline{\Omega}_0$, one can conclude that $b_{nn}(y) \geqslant k_0 > 0$ $(y \in \overline{\Omega}_0)$ (see (7.43)). It was shown above that $u_{y_p y_q} \in L_2(\Omega_0)$ $(p+q < 2n)$. Therefore,

$$u_{y_n y_n} = -\frac{1}{b_{nn}} \left(F + \sum_{p+q<2n} (b_{pq} u_{y_q})_{y_p} + (b_{nn})_{y_n} u_{y_n} \right) \in L_2(\Omega_0).$$

We see that $u(y) \in H^2(\Omega_0)$. Since the Sobolev space H^2 is invariant under diffeomorphisms of class C^2, we have $u(x) \in H^2(Q \cap B_{\delta/4}(x^0))$ and

$$\|u\|_{H^2(Q\cap B_{\delta/4}(x^0))} \leqslant k_5\big(\|f\|_{L_2(Q)} + \|u\|_{H^1(Q)}\big). \tag{7.54}$$

□

Remark 7.4. It is appropriate to mention why in Theorems 7.8 and 7.9 we did not restrict ourselves to the Poisson equation but considered more general elliptic equation (7.25) with variable coefficients from the beginning. There are several reasons for that.

First, we anyway arrive at elliptic equation (7.53) with variable coefficients after local straightening of the boundary in the proof of Theorem 7.9. Second, in what follows we will need the result established in Theorem 7.8 on the local smoothness of weak solutions (see the proof of Theorem 18.1).

Remark 7.5. The obtained result is generalized in a natural way. *Let $a_{ij}, a_i, a_0 \in C^\infty(\overline{Q})$. Suppose that $\partial Q \in C^{k+2}$ and $f \in H^k(Q)$. Then the weak solution to problem* (7.25), (7.2) *belongs to the space $H^{k+2}(Q)$. Moreover,*

$$\|u\|_{H^{k+2}(Q)} \leq c\big(\|f\|_{H^k(Q)} + \|u\|_{H^1(Q)}\big). \tag{7.55}$$

Combining Theorem 7.9, Theorem 7.1 and inequalities (7.39), (7.54), we come to the following statement.

Corollary 7.3. *Let $\partial Q \in C^2$ and $f \in L_2(Q)$. Then the weak solution $u \in \mathring{H}^1(Q)$ of problem* (7.1), (7.2) *belongs to the space $H^2(Q)$ and satisfies the inequality*

$$\|u\|_{H^2(Q)} \leq c_1\|f\|_{L_2(Q)}. \tag{7.56}$$

Below we make a number of important conclusions based on Theorem 7.9 and the Sobolev embedding Theorem 6.10, taking into account Remarks 6.6, 7.3 and 7.5.

Theorem 7.10. *Let $f \in L_2(Q) \cap H^{k+[n/2]-1}_{\mathrm{loc}}(Q)$ and $u \in \mathring{H}^1(Q)$ be a weak solution to problem* (7.1), (7.2). *Then $u \in C^k(Q)$.*

Theorem 7.11. *Let $\partial Q \in C^{k+[n/2]+1}$, $f \in H^{k+[n/2]-1}(Q)$ and $u \in \mathring{H}^1(Q)$ be a weak solution to problem* (7.1), (7.2). *Then $u \in C^k(\overline{Q})$.*

Theorem 7.12. *If $\partial Q \in C^{[n/2]+1}$ and $f \in H^{[n/2]-1}(Q) \cap H^{[n/2]+1}_{\mathrm{loc}}(Q)$, then the weak solution $u \in \mathring{H}^1(Q)$ to problem* (7.1), (7.2) *is the classical solution to this problem.*

Let us comment on the last result. Indeed, we have $f \in C(Q)$ and $u \in C^2(Q) \cap C(\overline{Q})$ under the conditions of the theorem. It only remains to add that the equality $-\Delta u = f$ holds everywhere in the domain Q, while the condition on the trace $u|_{\partial Q} = 0$ in the case of a continuous function u means that $u = 0$ at each point of ∂Q.

Theorem 7.13. *Any weak eigenfunction of the Dirichlet problem for the Laplace operator belongs to $C^\infty(Q)$. If, in addition, $\partial Q \in C^\infty$, then all weak eigenfunctions belong to $C^\infty(\overline{Q})$.*

It is enough to note that weak eigenfunctions can be considered as weak solutions to problem (7.1), (7.2) with the function $f = \lambda u \in \mathring{H}^1(Q)$ on right-hand side. Then iterations of Remark 7.5 (or Remark 7.3 in the case where no assumptions are made regarding the smoothness of the boundary ∂Q) show that $u \in H^k(Q)$ (or $u \in H^k_{\mathrm{loc}}(Q)$) for all positive integers k. But by virtue of the Sobolev embedding theorem, this is the same as $u \in C^\infty(\overline{Q})$ (or $u \in C^\infty(Q)$).

Remark 7.6. Theorem 7.10 means that a weak solution to problem (7.1), (7.2) coincides a. e. with a k-times continuously-differentiable function in Q, and the conclusions of Theorems 7.11–7.13 are interpreted similarly.

8 Neumann problem

The classical solution of the (nonhomogeneous) Neumann problem in a bounded domain $Q \subset \mathbb{R}^n$, $\partial Q \in C^1$ is a function $u \in C^2(Q) \cap C^1(\overline{Q})$ satisfying the Poisson equation

$$-\Delta u(x) = f(x) \quad (x \in Q) \tag{8.1}$$

and the boundary condition

$$\left.\frac{\partial u}{\partial \nu}\right|_{\partial Q} = \gamma(x) \quad (x \in \partial Q) \tag{8.2}$$

at each point of ∂Q, where the given functions f and γ are assumed continuous in Q and on ∂Q, respectively. In equation (8.2), $(\partial u/\partial \nu)|_{\partial Q}$ denotes the normal derivative of a function on the boundary. It can also be represented in the form

$$\left.\frac{\partial u}{\partial \nu}\right|_{\partial Q} = \nu \nabla u|_{\partial Q} = \sum_{i=1}^{n} \nu_i u_{x_i}|_{\partial Q},$$

where $\nu = \nu(x)$, $x \in \partial Q$, is the unit vector of the outer normal to the boundary, and ν_i is its ith coordinate.

If we additionally assume that $u \in C^2(\overline{Q})$, then we can multiply equation (8.1) by an arbitrary function $\overline{v} \in C^\infty(\overline{Q})$ and integrate the resulting equality over Q. Integrating by parts and taking relation (8.2) into account, we obtain

$$\int_Q \nabla u \nabla \overline{v}\, dx = \int_Q f\overline{v}\, dx + \int_{\partial Q} \gamma \overline{v}|_{\partial Q}\, dS. \tag{8.3}$$

At the same time, we make an important observation: in order to define all integrals in formula (8.3), it is enough to require only $u \in H^1(Q)$, $v \in H^1(Q)$, $f \in L_2(Q)$ and $\gamma \in L_2(\partial Q)$.

Definition 8.1. Let $f \in L_2(Q)$ and $\gamma \in L_2(\partial Q)$ be given functions. The function $u \in H^1(Q)$ is called a *weak solution* of problem (8.1), (8.2) if equation (8.3) holds for any $v \in H^1(Q)$.

In contrast to the Dirichlet problem, the boundary function γ in the weak formulation of the Neumann problem is an $L_2(\partial Q)$-function present in integral identity (8.3).

In a similar way, a weak eigenfunction of the Laplace operator with the Neumann condition is defined as a nontrivial weak solution to the problem

$$-\Delta u = \lambda u \quad (x \in Q),$$
$$\left.\frac{\partial u}{\partial \nu}\right|_{\partial Q} = 0.$$

Definition 8.2. A nonzero function $u \in H^1(Q)$ is called a *weak eigenfunction* of the Neumann problem for the Laplace operator corresponding to an *eigenvalue* λ if the integral identity

$$\int_Q \nabla u \nabla \bar{v}\, dx = \lambda \int_Q u\bar{v}\, dx \tag{8.4}$$

holds for all $v \in H^1(Q)$.

In order to study the Neumann problem, the same scheme is used as for the Dirichlet problem. A proof of the following statement completely repeats the reasoning used in the proof of Lemma 7.1.

Lemma 8.1. *There exists a unique bounded linear operator $A : L_2(Q) \to H^1(Q)$ such that*

$$(g, v)_{L_2(Q)} = (Ag, v)_{H^1(Q)} \quad (g \in L_2(Q), v \in H^1(Q)).$$

The restriction of A to $H^1(Q)$ is a positive compact operator on $H^1(Q)$.

Lemma 8.2. *A pair of functions $f \in L_2(Q)$, $\gamma \in L_2(\partial Q)$ uniquely determines a function $F \in H^1(Q)$ such that the equality*

$$(f, v)_{L_2(Q)} + (\gamma, v|_{\partial Q})_{L_2(\partial Q)} = (F, v)_{H^1(Q)} \tag{8.5}$$

holds for all functions $v \in H^1(Q)$.

To make sure of this, it is enough to note that the expression on the left-hand side of (8.5) represents again a continuous antilinear functional with respect to v on the space $H^1(Q)$. This easily follows from the Cauchy–Schwarz inequality and the trace theorem.

Theorem 8.1. *A weak solution to problem* (8.1), (8.2) *exists for those and only those functions f and γ that satisfy the relation*

$$\int_Q f\, dx + \int_{\partial Q} \gamma\, dS = 0. \tag{8.6}$$

If exists, this solution is determined up to an arbitrary additive constant.

Proof. Using Lemmas 8.1 and 8.2, we represent identity (8.3) as follows:

$$(u,v)_{H^1(Q)} = (u,v)_{L_2(Q)} + (f,v)_{L_2(Q)} + (\gamma, v|_{\partial Q})_{L_2(\partial Q)}$$

or

$$(u,v)_{H^1(Q)} = (Au,v)_{H^1(Q)} + (F,v)_{H^1(Q)}. \tag{8.7}$$

Therefore, we arrive at the equivalent operator equation $u - Au = F$ with a positive compact operator A in the space $H^1(Q)$. According to Fredholm theory, the homogeneous equation $u = Au$ is considered first. From Lemma 8.1 and equation (8.7), it follows that $\|u\|_{H^1(Q)} = \|u\|_{L_2(Q)}$ or $\int_Q |\nabla u|^2\,dx = 0$. This proves that the solution space of the homogeneous problem consists of constants. But then the nonhomogeneous equation $u - Au = F$ is solvable if and only if the function F is orthogonal to constants in the space $H^1(Q)$, the last condition is (8.6) due to (8.5). □

Theorem 8.2. *The eigenvalues of the Neumann problem for the negative Laplace operator are real, isolated, have finite multiplicity and with an appropriate enumeration taking into account their multiplicities, can be written as the sequence*

$$\lambda_1 = 0 < \lambda_2 \leqslant \cdots \leqslant \lambda_s \leqslant \ldots, \quad \lambda_s \to +\infty, \quad s \to \infty.$$

There exists an orthonormal basis in the space $L_2(Q)$ consisting of eigenfunctions

$$u_1 = |Q|^{-1/2}, u_2, \ldots, u_s, \ldots .$$

Moreover, the functions

$$u_1, \frac{u_2}{\sqrt{\lambda_2 + 1}}, \ldots, \frac{u_s}{\sqrt{\lambda_s + 1}}, \ldots$$

form an orthonormal basis in the space $H^1(Q)$.

Proof. It follows from (8.4) that

$$\lambda = \frac{\int_Q |\nabla u|^2\,dx}{\int_Q |u|^2\,dx} \geqslant 0,$$

where u is a weak eigenfunction corresponding to an eigenvalue λ of the Laplace operator with the Neumann condition, while Theorem 8.1 states that $\lambda = 0$ is an eigenvalue, and the corresponding eigenfunction is a constant.

Let us represent identity (8.4) in the form

$$(u,v)_{H^1(Q)} = (\lambda + 1)(u,v)_{L_2(Q)} = (\lambda + 1)(Au,v)_{H^1(Q)} \quad (v \in H^1(Q)).$$

Thus, we reduce the eigenvalue problem for the Laplace operator with the Neumann condition to the eigenvalue problem

$$Au = \mu u, \quad \mu = 1/(\lambda + 1)$$

for a positive compact operator $A : H^1(Q) \to H^1(Q)$. After this, the reasoning completely repeats the proof of Theorem 7.3. □

Remark 8.1. The Sobolev space $H^1(Q)$ can be equivalently described using the Fourier coefficients of a function over the system of eigenfunctions of the Neumann problem for the Laplace operator.

Note that any function $f \in L_2(Q)$ can be expanded into a Fourier series over the orthonormal basis $\{u_s,\ s = 1, 2, \dots\}$ in the space $L_2(Q)$, i. e.,

$$f(x) = \sum_{s=1}^{\infty} f_s u_s(x), \quad f_s = (f, u_s)_{L_2(Q)}. \tag{8.8}$$

Since all functions u_s are pairwise orthogonal in $H^1(Q)$, while $\|u_s\|^2_{H^1(Q)} = \lambda_s + 1$, we have the formula

$$\left\| \sum_{s=m+1}^{m+p} f_s u_s \right\|^2_{H^1(Q)} = \sum_{s=m+1}^{m+p} (\lambda_s + 1)|f_s|^2$$

for partial sums of series (8.8). Therefore, the convergence of the number series

$$\sum_{s=1}^{\infty} \lambda_s |f_s|^2 < \infty \tag{8.9}$$

implies the convergence of series (8.8) in the space $H^1(Q)$. In other words, (8.9) guarantees that $f \in H^1(Q)$.

Conversely, let a function f belong to $H^1(Q)$. Then, along with expansion (8.8), a similar expansion can be written with respect to the orthonormal basis $u_s/\sqrt{\lambda_s + 1}$, $s = 1, 2, \dots$, in the space $H^1(Q)$,

$$f(x) = \sum_{s=1}^{\infty} f_s' \frac{u_s(x)}{\sqrt{\lambda_s + 1}}, \quad f_s' = \left(f, \frac{u_s}{\sqrt{\lambda_s + 1}}\right)_{H^1(Q)}.$$

It follows from the definition of u_s that $(f, u_s)_{H^1(Q)} = (\lambda_s + 1)(f, u_s)_{L_2(Q)}$, i. e., $f_s' = \sqrt{\lambda_s + 1} f_s$. Therefore, Parseval's equality in $H^1(Q)$ takes the form

$$\|f\|^2_{H^1(Q)} = \sum_{s=1}^{\infty} |f_s'|^2 = \sum_{s=1}^{\infty} (\lambda_s + 1)|f_s|^2.$$

Hence if $f \in H^1(Q)$, then (8.9) holds.

Remark 8.2. The condition $\int_Q u\,dx = 0$ natural from a physical point of view is often added to the formulation of the Neumann problem. This condition identifies the orthogonal complement to the one-dimensional subspace of $H^1(Q)$ consisting of constant functions in the domain. We denote this orthogonal complement by $\widetilde{H}^1(Q)$. It is easy to see that the weak solution u from the space $\widetilde{H}^1(Q)$ is unique and, if exists, depends continuously on the functions f and γ, i. e.,

$$\|u\|_{H^1(Q)} \leqslant c\left(\|f\|_{L_2(Q)} + \|\gamma\|_{L_2(\partial Q)}\right).$$

A weak solution of the Neumann problem is often understood as an element u of the space $\widetilde{H}^1(Q)$.

As in the case of the Dirichlet problem, the weak solution of the Neumann problem can be represented as a Fourier series. Assuming that condition (8.6) is satisfied, we look for a solution u to problem (8.1), (8.2) in the subspace $\widetilde{H}^1(Q)$,

$$u(x) = \sum_{s=2}^{\infty} a_s \frac{u_s(x)}{\sqrt{\lambda_s + 1}}, \quad a_s = \left(u, \frac{u_s}{\sqrt{\lambda_s + 1}}\right)_{H^1(Q)} \qquad (s = 2, 3, \ldots).$$

In order to obtain a_s, substitute u_s for v in (8.3). Then we get

$$a_s\sqrt{\lambda_s + 1} = (u, u_s)_{L_2(Q)} + (f, u_s)_{L_2(Q)} + (\gamma, u_s|_{\partial Q})_{L_2(\partial Q)}.$$

On the other hand, it follows from the definition of u_s that

$$a_s\sqrt{\lambda_s + 1} = (u, u_s)_{H^1(Q)} = (\lambda_s + 1)(u, u_s)_{L_2(Q)}.$$

Combining these relations, we get

$$u(x) = \sum_{s=2}^{\infty} \frac{(f, u_s)_{L_2(Q)} + (\gamma, u_s|_{\partial Q})_{L_2(\partial Q)}}{\lambda_s} u_s(x).$$

Let us now turn to the question of smoothness of weak solutions of the Neumann problem. First, narrowing down the class of test functions v to $\mathring{H}^1(Q)$ in (8.3), we see that a weak solution $u(x)$ of problem (8.1), (8.2) satisfies the same identity

$$\int_Q \nabla u \nabla \bar{v}\,dx = \int_Q f\bar{v}\,dx \quad (v \in \mathring{H}^1(Q)),$$

as in the case of the Dirichlet problem. Therefore, we already have the internal smoothness of weak solutions of the Neumann problem, i. e., $u \in H^{k+2}_{\text{loc}}(Q)$ for $f \in L_2(Q) \cap H^k_{\text{loc}}(Q)$ (see Remark 7.3). In addition, the weak solution satisfies the equation $-\Delta u = f$ a. e. in Q.

Studying smoothness of weak solutions of problem (8.1), (8.2) in the entire domain, we consider the case of the homogeneous boundary condition, $\gamma = 0$. In this case, a weak solution of problem (8.1), (8.2) is a function $u \in H^1(Q)$ that satisfies the similar identity

$$\int_Q \nabla u \nabla \overline{v}\, dx = \int_Q f\overline{v}\, dx \quad (v \in H^1(Q)), \tag{8.10}$$

but now for all v from $H^1(Q)$. The proof of Theorem 7.9, based on the approximation of weak derivatives by finite differences, works also in the current situation with the space $\mathring{H}^1(Q)$ replaced by the space $H^1(Q)$. Note that Theorem 5.6 does not require the trace of the function on the flat part of the boundary to be equal to zero.

Theorem 8.3. *If $\partial Q \in C^{k+2}$ and $f \in H^k(Q)$, then a weak solution $u(x)$ of problem* (8.1), (8.2) *with $\gamma = 0$ belongs to the space $H^{k+2}(Q)$ and, under the additional condition $\int_Q u\, dx = 0$, satisfies the inequality*

$$\|u\|_{H^{k+2}(Q)} \leqslant c\|f\|_{H^k(Q)},$$

in which the constant $c > 0$ does not depend on f.

For $k = 0$, the weak solution satisfies boundary condition (8.2), *where*

$$\left.\frac{\partial u}{\partial \nu}\right|_{\partial Q} = \nu\nabla u|_{\partial Q}, \quad \nabla u|_{\partial Q} = (u_{x_1}|_{\partial Q}, \dots, u_{x_n}|_{\partial Q}), \quad u_{x_j} \in H^1(Q).$$

In order to check the second statement of the theorem, integrate (8.10) by parts. We have

$$\int_{\partial Q} \nu\nabla u|_{\partial Q}\overline{v}|_{\partial Q}\, dx - \int_Q \Delta u\overline{v}\, dx = \int_Q f\overline{v}\, dx.$$

Taking into account that $\Delta u + f = 0$ a. e. in Q and also that the set of traces on the boundary of all functions from $H^1(Q)$ is dense in $L_2(\partial Q)$, we get $\nu\nabla u|_{\partial Q} = 0$.

Combining Theorem 8.3 with the Sobolev embedding theorem (Theorem 6.10), we get the following result.

Theorem 8.4. *If $\partial Q \in C^{[n/2]+2}$ and $f \in H^{[n/2]+1}(Q)$, then a weak solution to problem* (8.1), (8.2) *with $\gamma = 0$ is its classical solution.*

9 Classical solutions to the Poisson equation

Integral representation of functions

Let us recall a special version of the formula for integration by parts,

$$\int_Q (v\Delta u - u\Delta v)\, dx = \int_{\partial Q} \left(v\frac{\partial u}{\partial \nu} - u\frac{\partial v}{\partial \nu}\right) dS \quad (u, v \in C^2(\overline{Q})). \tag{9.1}$$

Here, Q is a bounded domain in $\mathbb{R}^n$ with boundary of class C^1. Formula (9.1) is called *Green's formula.* In order to obtain (9.1), subtract the equality

$$\int_Q u\Delta v\,dx = \int_{\partial Q} u\frac{\partial v}{\partial \nu}\,dS - \int \nabla u \nabla v\,dx$$

from the similar equality

$$\int_Q v\Delta u\,dx = \int_{\partial Q} v\frac{\partial u}{\partial \nu}\,dS - \int \nabla u \nabla v\,dx.$$

Let us introduce a function $U \in C^\infty(\mathbb{R}^n \setminus \{0\})$ according to the formula

$$U(x) = \begin{cases} \frac{1}{(2-n)\sigma_n}|x|^{2-n}, & n \geqslant 3, \\ \frac{1}{2\pi}\ln|x|, & n = 2, \end{cases} \tag{9.2}$$

where σ_n is the surface area of the unit sphere S^{n-1} in $\mathbb{R}^n$. Direct verification shows that U satisfies the Laplace equation $\Delta U(x) = 0$ everywhere except for the origin. At the origin, U has an integrable singularity. This function plays an essential role in the study of solutions to the Laplace and Poisson equations. We will also need the obvious relation

$$\frac{\partial U}{\partial |x|} = \frac{1}{\sigma_n |x|^{n-1}}, \quad n \geqslant 2.$$

Definition 9.1. Function (9.2) is called the *fundamental solution* of the Laplace equation.

Theorem 9.1. *Let $Q \subset \mathbb{R}^n$ be a bounded domain, $\partial Q \in C^1$. Then any function $f \in C^2(\overline{Q})$ satisfies the representation*

$$f(x) = \int_Q U(x-\xi)\Delta f(\xi)\,d\xi + \\ + \int_{\partial Q}\left(\frac{\partial U(x-\xi)}{\partial \nu_\xi}f(\xi) - U(x-\xi)\frac{\partial f(\xi)}{\partial \nu_\xi}\right)dS_\xi \quad (x \in Q). \tag{9.3}$$

Proof. We assume without loss of generality that $n \geqslant 3$. Let us take an arbitrary point $x \in Q$ and consider a number $\varepsilon > 0$ such that $\overline{B_\varepsilon(x)} \subset Q$. We then move on to the multiconnected domain $\Omega_\varepsilon = Q\setminus\overline{B_\varepsilon(x)}$ with smooth boundary, having $f, U(x-\cdot) \in C^2(\overline{\Omega}_\varepsilon)$. Applying Green's formula to the pair $f, U(x-\cdot)$ in the domain Ω_ε and taking into account that $\Delta_\xi U(x-\xi) = 0$ in Ω_ε, we obtain

$$\int_{\Omega_\varepsilon} U(x-\xi)\Delta f(\xi)\,d\xi = \int_{\partial Q}\left(U(x-\xi)\frac{\partial f(\xi)}{\partial \nu_\xi} - f(\xi)\frac{\partial U(x-\xi)}{\partial \nu_\xi}\right)dS_\xi +$$

$$+\int\limits_{|\xi-x|=\varepsilon}\left(U(x-\xi)\frac{\partial f(\xi)}{\partial\nu_\xi}-f(\xi)\frac{\partial U(x-\xi)}{\partial\nu_\xi}\right)dS_\xi$$

(the direction of the normal to the sphere $|\xi - x| = \varepsilon$ is external with respect to the domain Ω_ε; the normal is internal with respect to the cut-out ball). Let us estimate the contribution of the second integral to the right-hand side of this equality.

If $C_f = \max_{\xi\in\overline{Q}}|\nabla f(\xi)|$, then it is easy to see that $|f(x)-f(y)| \leqslant C_f|x-y|$ when the segment $[x,y]$ lies entirely in Q.

Denote

$$I_{1\varepsilon}=\int\limits_{|\xi-x|=\varepsilon}U(x-\xi)\frac{\partial f(\xi)}{\partial\nu_\xi}\,dS_\xi,\quad I_{2\varepsilon}=-\int\limits_{|\xi-x|=\varepsilon}f(\xi)\frac{\partial U(x-\xi)}{\partial\nu_\xi}\,dS.$$

Since

$$I_{1\varepsilon}=\frac{1}{(2-n)\sigma_n\varepsilon^{n-2}}\int\limits_{|\xi-x|=\varepsilon}\frac{\partial f(\xi)}{\partial\nu_\xi}\,dS_\xi,$$

we have

$$|I_{1\varepsilon}|\leqslant\frac{1}{(n-2)\sigma_n\varepsilon^{n-2}}\int\limits_{|\xi-x|=\varepsilon}|\nabla f(\xi)|\,dS_\xi\leqslant\frac{C_f\sigma_n\varepsilon^{n-1}}{(n-2)\sigma_n\varepsilon^{n-2}}=\frac{C_f\varepsilon}{n-2}.$$

Therefore, $I_{1\varepsilon}\to 0$ as $\varepsilon\to 0$. Taking the relations

$$\left.\frac{\partial U(x-\xi)}{\partial\nu_\xi}\right|_{|\xi-x|=\varepsilon}=-\left.\frac{\partial U(x-\xi)}{\partial|x-\xi|}\right|_{|\xi-x|=\varepsilon}=-\frac{1}{\sigma_n\varepsilon^{n-1}}$$

into account, we can write

$$I_{2\varepsilon}=\frac{1}{\sigma_n\varepsilon^{n-1}}\int\limits_{|\xi-x|=\varepsilon}f(\xi)\,dS_\xi=f(x)+\frac{1}{\sigma_n\varepsilon^{n-1}}\int\limits_{|\xi-x|=\varepsilon}(f(\xi)-f(x))\,dS_\xi.$$

Therefore,

$$|I_{2\varepsilon}-f(x)|\leqslant\frac{1}{\sigma_n\varepsilon^{n-1}}\int\limits_{|\xi-x|=\varepsilon}|f(\xi)-f(x)|\,dS_\xi\leqslant\frac{C_f\varepsilon}{\sigma_n\varepsilon^{n-1}}\int\limits_{|\xi-x|=\varepsilon}dS_\xi=C_f\varepsilon,$$

i. e., $I_{2\varepsilon}\to f(x)$ as $\varepsilon\to 0$.

Collecting all the calculations and noting that

$$\int\limits_{\Omega_\varepsilon}U(x-\xi)\Delta f(\xi)\,d\xi\to\int\limits_{Q}U(x-\xi)\Delta f(\xi)\,d\xi,\quad\varepsilon\to 0,$$

we arrive at the desired result. □

Remark 9.1. Formula (9.3) is not suitable to solve directly the Dirichlet or Neumann problems for the Poisson equation, since it requires specifying both $f|_{\partial Q}$ and $(\partial f/\partial \nu)|_{\partial Q}$ simultaneously. Nevertheless, this formula turns out to be very useful in the further study of qualitative properties of classical solutions of boundary-value problems for the Poisson equation.

Mean value theorems for harmonic functions

Definition 9.2. A function $u(x)$ is called *harmonic* in an open set $Q \subset \mathbb{R}^n$ if $u \in C^2(Q)$ and $\Delta u = 0$ in Q.

Lemma 9.1. *Let Ω be a bounded domain with boundary of class C^1, $\overline{\Omega} \subset Q$ and $u(x)$ be a harmonic function in Q. Then*

$$\int_{\partial\Omega} \frac{\partial u}{\partial \nu}\, dS = 0.$$

This statement is a simple consequence of the integration by parts formula,

$$\int_{\partial\Omega} \frac{\partial u}{\partial \nu}\, dS = \int_{\Omega} \Delta u\, dx = 0.$$

Theorem 9.2. *Let $Q \subset \mathbb{R}^n$ be an open set, $\overline{B_d(x)} \subset Q$ and $u(x)$ be a harmonic function in Q. Then*

$$u(x) = \frac{1}{\sigma_n d^{n-1}} \int_{|\xi - x| = d} u(\xi)\, dS_\xi,$$

i. e., the value $u(x)$ of a harmonic function at the center of a sphere is equal to the average value of this function on the sphere.

Proof. Apply formula (9.3) to a function $u \in C^2(\overline{B_d(x)})$, $\Delta u = 0$. Taking Lemma 9.1 into account, we obtain the desired relation since

$$\begin{aligned} u(x) &= \int_{|\xi - x| = d} \left(\frac{\partial U(x-\xi)}{\partial \nu_\xi} u(\xi) - U(x-\xi) \frac{\partial u(\xi)}{\partial \nu_\xi} \right) dS_\xi = \\ &= \frac{1}{\sigma_n d^{n-1}} \int_{|\xi - x| = d} u(\xi)\, dS_\xi + \frac{1}{(n-2)\sigma_n d^{n-2}} \int_{|\xi - x| = d} \frac{\partial u}{\partial \nu_\xi}\, dS_\xi = \\ &= \frac{1}{\sigma_n d^{n-1}} \int_{|\xi - x| = d} u(\xi)\, dS_\xi. \end{aligned}$$

□

Theorem 9.3. *Under the conditions of Theorem 9.2, the equality*

$$u(x) = \frac{n}{\sigma_n d^n} \int\limits_{|\xi - x| < d} u(\xi)\, d\xi$$

is also valid, i. e., the value $u(x)$ of a harmonic function at the center of a ball is also equal to the average value of this function over the ball.

Proof. Apply Theorem 9.2 for all $B_\rho(x)$, $0 < \rho < d$,

$$u(x)\sigma_n \rho^{n-1} = \int\limits_{|\xi - x| = \rho} u(\xi)\, dS_\xi \quad (0 < \rho < d).$$

Integrating this equality over ρ, we obtain

$$u(x)\sigma_n \int\limits_0^d \rho^{n-1} d\rho = \int\limits_0^d d\rho \int\limits_{|\xi - x| = \rho} u(\xi)\, dS_\xi$$

or

$$\frac{u(x)\sigma_n d^n}{n} = \int\limits_{|\xi - x| < d} u(\xi)\, d\xi,$$

where $\sigma_n d^n / n$ denotes the volume of the ball $B_d(x)$. □

Maximum principle

The next important property of harmonic functions based on Theorem 9.3 is called the *maximum principle.*

Theorem 9.4. *Let Q be a bounded domain in $\mathbb{R}^n$ and a real-valued function $u(x)$ be harmonic in Q and continuous in $\overline{Q}$. Then either this function is constant in Q or*

$$\min_{x \in \partial Q} u(x) < u(y) < \max_{x \in \partial Q} u(x) \quad (y \in Q). \tag{9.4}$$

Proof. Let $M = \max_{x \in \overline{Q}} u(x)$. Suppose that there exists a point $x^0 \in Q$ such that $u(x^0) = M$. Let us show that $u \equiv M$ in this case.

Having fixed an arbitrary point $y \in Q$, we connect it to the point x^0 by a polygonal line L without self-intersections, consisting of a finite number of links and lying entirely in Q. Let us denote $d = \rho(L, \partial Q) > 0$. Cover L with a finite number of balls $B_i = \{|x - x^i| < d/2\}$ with centers at points $x^0, x^i \in L \cap B_{i-1}$, $i = 1, \dots, N$, where $x^N = y$.

Theorem 9.3 implies

$$u(x^0) = \frac{n}{\sigma_n (d/2)^n} \int_{B_0} u(x)\,dx,$$

i. e.,

$$\int_{B_0} (u(x^0) - u(x))\,dx = 0.$$

By assumption, the integrand function $u(x^0) - u(x)$ is nonnegative and continuous. Therefore, $u(x) = u(x^0) = M$ in B_0. In particular, $u(x^1) = M$. Similarly,

$$\int_{B_1} (u(x^1) - u(x))\,dx = 0$$

by Theorem 9.3. Hence $u(x^2) = M$, and iterating, we get $u(y) = M$. Therefore, $u(x) \equiv M$.

Applying the same reasoning to the function $-u(x)$, we prove that either u is a constant or the left-hand side of inequality (9.4) holds. □

Recall that a function $u(x)$ is a classical solution to the Dirichlet problem

$$-\Delta u = f \quad (x \in Q), \tag{9.5}$$

$$u|_{\partial Q} = \gamma \quad (x \in \partial Q) \tag{9.6}$$

if u belongs to $C^2(Q) \cap C(\overline{Q})$ and satisfies relations (9.5) and (9.6) at each point of Q and ∂Q, respectively, where f and γ are continuous functions.

It immediately follows from the maximum principle that the classical solution to problem (9.5), (9.6) is unique.

Theorem 9.5. *The Dirichlet problem for the Poisson equation has at most one classical solution.*

Proof. Suppose that both functions $u_1(x)$ and $u_2(x)$ are classical solutions to problem (9.5), (9.6). But then the function $u(x) = u_1(x) - u_2(x)$ is harmonic in Q, continuous in $\overline{Q}$ and vanishes on ∂Q. Therefore, Theorem 9.4 states that $u(x) \equiv 0$, i. e., $u_1(x) \equiv u_2(x)$ in Q. □

Remark 9.2. The question of existence of the classical solution to the Dirichlet problem is much more difficult. For example, problem (9.5), (9.6) may not have the classical solution for a function f continuous in $\overline{Q}$. On the other hand, conditions $\partial Q \in C^2$ and $f \in C^1(\overline{Q})$ already guarantee the existence of the classical solution. One can also prove a stronger statement about the existence of the classical solution under the assumption that the function f is Hölder continuous in $\overline{Q}$. The latter means the existence of numbers

$0 < \alpha < 1$ and $M > 0$ such that the inequality

$$|f(x) - f(y)| \leqslant M|x - y|^{\alpha}$$

holds for all $x, y \in Q$. We do not consider this issue in the textbook.

4 Mixed problems and the Cauchy problem for hyperbolic and parabolic equations

10 Mixed problems for the wave equation

We focus on the first mixed problem in this section. The other mixed problems, i.e, the second and third, are studied in quite a similar way. Function spaces throughout the chapter are assumed to be real.

Uniqueness of weak solution

Let Q be a bounded domain in $\mathbb{R}^n_x$ and $T > 0$. Consider the cylinder $\Omega_T = Q\times(0,T) \subset \mathbb{R}^{n+1}_{x,t}$. Denote its lateral surface by $\Gamma_T = \partial Q \times (0,T)$, while Q_τ is the cross-section of the cylinder by the hyperplane $\{t = \tau\}$.

Recall that the first mixed problem for the wave equation consists of finding a solution to the wave equation

$$u_{tt} - \Delta u = f(x,t) \quad ((x,t) \in \Omega_T), \tag{10.1}$$

satisfying the initial conditions

$$u|_{t=0} = \varphi(x) \quad (x \in Q), \tag{10.2}$$

$$u_t|_{t=0} = \psi(x) \quad (x \in Q), \tag{10.3}$$

as well as the boundary condition

$$u|_{\Gamma_T} = \gamma(x,t) \quad ((x,t) \in \Gamma_T). \tag{10.4}$$

By changing the unknown function, problem (10.4) with the nonhomogeneous boundary condition can be reduced to the problem with the homogeneous boundary condition

$$u|_{\Gamma_T} = 0. \tag{10.5}$$

For this reason, we restrict ourselves to problem (10.1), (10.2), (10.3) and (10.5), which we call the *mixed problem* for brevity. We assume that the boundary ∂Q belongs to the class C^1.

A function $u \in C^2(\Omega_T) \cap C^1(\Omega_T \cup \overline{Q}_0) \cap C(\Omega_T \cup \Gamma_T \cup \overline{Q}_0)$ is said to be a classical solution to problem (10.1)–(10.3) and (10.5) if it satisfies equation (10.1) in the cylinder Ω_T, initial conditions (10.2) and (10.3) on its lower base Q_0, and boundary condition (10.5) on the lateral surface Γ_T.

Suppose that $u \in C^2(\overline{\Omega}_T)$ is a classical solution to mixed problem (10.1)–(10.3), (10.5). In order to introduce the concept of weak solution, we integrate equation (10.1) multi-

https://doi.org/10.1515/9783112229637-004

plied by an arbitrary function $v \in H^1(\Omega_T)$ such that $v|_{\Gamma_T} = 0$ and $v|_{Q_T} = 0$, over Ω_T,

$$\int_{\Omega_T} (u_{tt}v - \Delta u v)\,dxdt = \int_{\Omega_T} fv\,dxdt.$$

Integrating by parts on the left-hand side of this equality and taking into account the conditions imposed on the function v, we obtain

$$\int_{\Omega_T} (\nabla u \nabla v - u_t v_t)\,dxdt = \int_{\Omega_T} fv\,dxdt + \int_{Q} \psi v|_{t=0}\,dx. \tag{10.6}$$

We introduce the notation $\widetilde{H}^1(\Omega_T) = \{v \in H^1(\Omega_T) : v|_{\Gamma_T} = 0, v|_{Q_T} = 0\}$.

Definition 10.1. Suppose that $f \in L_2(\Omega_T)$ and $\varphi, \psi \in L_2(Q)$. A function $u \in H^1(\Omega_T)$ is called *a weak solution* of mixed problem (10.1)–(10.3) and (10.5) if it satisfies integral identity (10.6) for all functions $v \in \widetilde{H}^1(\Omega_T)$, and conditions (10.2) and (10.5) understood in the sense of the trace of an H^1-function on a smooth n-dimensional manifold.

Theorem 10.1. *Mixed problem* (10.1)–(10.3), (10.5) *has at most one weak solution.*

Proof. Due to the linearity of the problem under consideration, it suffices to verify that the corresponding homogeneous problem for $f = 0$ and $\varphi = \psi = 0$ has the only trivial solution, i. e., if a function $u \in H^1(\Omega_T)$ satisfies the identity

$$\int_{\Omega_T} (\nabla u \nabla v - u_t v_t)\,dxdt = 0 \tag{10.7}$$

for all $v \in \widetilde{H}^1(\Omega_T)$ and is such that $u|_{t=0} = 0$ and $u|_{\Gamma_T} = 0$, then $u(x,t) = 0$ a. e. in Ω_T. For this purpose, fix an arbitrary $\tau \in (0,T)$ and choose a test function v in (10.7) as follows:

$$v(x,t) = \begin{cases} \int_t^\tau u(x,\theta)\,d\theta, & 0 < t < \tau, \\ 0, & \tau < t < T. \end{cases}$$

It is easy to see that $v \in \widetilde{H}^1(\Omega_T)$,

$$v_t(x,t) = \begin{cases} -u(x,t), & 0 < t < \tau, \\ 0, & \tau < t < T \end{cases}$$

and

$$\nabla v(x,t) = \begin{cases} \int_t^\tau \nabla u(x,\theta)\,d\theta, & 0 < t < \tau, \\ 0, & \tau < t < T. \end{cases}$$

By Fubini's theorem, identity (10.7) takes the form

$$\int_Q dx \int_0^\tau \nabla u(x,t)\,dt \int_t^\tau \nabla u(x,\theta)\,d\theta + \int_Q dx \int_0^\tau u_t(x,t)u(x,t)\,dt = 0. \tag{10.8}$$

Integrating by parts in the second term, we have

$$I_2 = \int_{\Omega_\tau} u_t u\,dxdt = \int_Q (u|_{t=\tau})^2 dx - \int_{\Omega_\tau} uu_t\,dxdt,$$

i. e.,

$$I_2 = \frac{1}{2}\int_Q (u|_{t=\tau})^2 dx \geqslant 0.$$

For the inner integral in the first term on the left-hand side in (10.8), we use again Fubini's theorem to get

$$\begin{aligned} I_1 &= \int_Q dx \int_0^\tau \nabla u(x,t)\,dt \int_t^\tau \nabla u(x,\theta)\,d\theta = \\ &= \int_Q dx \int_0^\tau \nabla u(x,\theta)\,d\theta \int_0^\theta \nabla u(x,t)\,dt = \int_Q dx \int_0^\tau \nabla u(x,\theta)\,d\theta \int_0^\tau \nabla u(x,t)\,dt - \\ &\quad - \int_Q dx \int_0^\tau \nabla u(x,\theta)\,d\theta \int_\theta^\tau \nabla u(x,t)\,dt = \int_Q \left| \int_0^\tau \nabla u(x,t)\,dt \right|^2 dx - I_1, \end{aligned}$$

i. e.,

$$I_1 = \frac{1}{2}\int_Q \left| \int_0^\tau \nabla u(x,t)\,dt \right|^2 dx \geqslant 0.$$

It follows now from (10.8) that $\int_Q (u|_{t=\tau})^2 dx = 0$. Since $\tau \in (0,T)$ is arbitrary, we conclude by Remark 6.5 that $u = 0$ in Ω_T. □

Fourier method for the wave equation

The solution of mixed problem (10.1)–(10.3) and (10.5) by the Fourier method is based on the spectral properties of the Laplace operator with the Dirichlet condition in Q. Recall that the eigenfunctions $e_1, e_2, \ldots$ of the problem

$$-\Delta e(x) = \lambda e(x) \quad (x \in Q), \tag{10.9}$$

$$e|_{\partial Q} = 0 \tag{10.10}$$

form an orthonormal basis in the space $L_2(Q)$, and the corresponding eigenvalues $\lambda_1, \lambda_2, \dots$ are positive and are represented as the sequence $0 < \lambda_1 < \lambda_2 \leqslant \lambda_3 \leqslant \cdots \leqslant \lambda_k \leqslant \dots \to \infty$. Eigenfunctions can be considered real-valued since the real and imaginary parts of a complex-valued eigenfunction are also eigenfunctions corresponding to the same eigenvalue.

Let us begin with a formal description of the method. After this, based on the given algorithm, a rigorous proof of the existence of a weak solution to the mixed problem is given. According to the Fourier method, the functions $\varphi(x)$, $\psi(x)$ and $f(x,t)$ are expanded in $L_2(Q)$ into the Fourier series in the basis $\{e_k\}_{k=1}^{\infty}$. Namely,

$$\varphi(x) = \sum_{k=1}^{\infty} \varphi_k e_k(x), \quad \varphi_k = (\varphi, e_k)_{L_2(Q)},$$

$$\psi(x) = \sum_{k=1}^{\infty} \psi_k e_k(x), \quad \psi_k = (\psi, e_k)_{L_2(Q)},$$

$$f(x,t) = \sum_{k=1}^{\infty} f_k(t) e_k(x), \quad f_k(t) = (f(\cdot,t), e_k)_{L_2(Q)}.$$

The solution to the mixed problem is also sought in the form of a Fourier series

$$u(x,t) = \sum_{k=0}^{\infty} u_k(t) e_k(x), \tag{10.11}$$

where coefficients $u_k(t)$ are to be determined.

Putting series (10.11) in the wave equation and taking into account that $\Delta e_k = -\lambda_k e_k$, we obtain the equality

$$\sum_{k=1}^{\infty} u_k''(t) e_k(x) + \sum_{k=1}^{\infty} \lambda_k u_k(t) e_k(x) = \sum_{k=1}^{\infty} f_k(t) e_k(x)$$

resulting in the corresponding equations

$$u_k''(t) + \lambda_k u_k(t) = f_k(t) \quad (0 < t < T;\ k = 1, 2, \dots) \tag{10.12}$$

for the coefficients $u_k(t)$.

Similarly, the initial conditions

$$u|_{t=0} = \sum_{k=0}^{\infty} u_k(0) e_k(x) = \sum_{k=1}^{\infty} \varphi_k e_k(x)$$

and

$$u_t|_{t=0} = \sum_{k=0}^{\infty} u_k'(0) e_k(x) = \sum_{k=1}^{\infty} \psi_k e_k(x)$$

lead to the equalities

$$u_k(0) = \varphi_k, \quad u_k'(0) = \psi_k \quad (k = 1, 2, \ldots). \tag{10.13}$$

Note that the boundary condition $u|_{\Gamma_T} = 0$ is already satisfied since the basis functions $e_k(x)$ are such that $e_k|_{\partial Q} = 0$.

Obtained initial value problems (10.12), (10.13) can be solved explicitly. The function $a_k \cos \sqrt{\lambda_k} t + b_k \sin \sqrt{\lambda_k} t$ depending on arbitrary real parameters a_k and b_k is the general solution of the corresponding homogeneous equation $u_k'' + \lambda_k u_k = 0$. In accordance with the method of variation of parameters, a particular solution of nonhomogeneous equation (10.12) can be represented as $a_k(t) \cos \sqrt{\lambda_k} t + b_k(t) \sin \sqrt{\lambda_k} t$, where the functions $a_k(t)$ and $b_k(t)$ are determined from the system

$$\begin{cases} a_k'(t) \cos \sqrt{\lambda_k} t + b_k'(t) \sin \sqrt{\lambda_k} t = 0, \\ \sqrt{\lambda_k}(-a_k'(t) \sin \sqrt{\lambda_k} t + b_k'(t) \cos \sqrt{\lambda_k} t) = f_k(t). \end{cases}$$

Carrying out obvious calculations

$$a_k(t) = -\frac{1}{\sqrt{\lambda_k}} \int_0^t f_k(\tau) \sin \sqrt{\lambda_k} \tau \, d\tau, \quad b_k(t) = \frac{1}{\sqrt{\lambda_k}} \int_0^t f_k(\tau) \cos \sqrt{\lambda_k} \tau \, d\tau,$$

we see that the function

$$-\frac{\cos \sqrt{\lambda_k} t}{\sqrt{\lambda_k}} \int_0^t f_k(\tau) \sin \sqrt{\lambda_k} \tau \, d\tau + \frac{\sin \sqrt{\lambda_k} t}{\sqrt{\lambda_k}} \int_0^t f_k(\tau) \cos \sqrt{\lambda_k} \tau \, d\tau =$$
$$= \frac{1}{\sqrt{\lambda_k}} \int_0^t f_k(\tau) \sin \sqrt{\lambda_k}(t - \tau) \, d\tau$$

is a particular solution to equation (10.12), and the function

$$u_k(t) = a_k \cos \sqrt{\lambda_k} t + b_k \sin \sqrt{\lambda_k} t + \frac{1}{\sqrt{\lambda_k}} \int_0^t f_k(\tau) \sin \sqrt{\lambda_k}(t - \tau) \, d\tau$$

represents the general solution to equation (10.12). Finally, after using relations (10.13), we obtain the solution to problem (10.12), (10.13):

$$u_k(t) = \varphi_k \cos \sqrt{\lambda_k} t + \frac{\psi_k}{\sqrt{\lambda_k}} \sin \sqrt{\lambda_k} t + \\ + \frac{1}{\sqrt{\lambda_k}} \int_0^t f_k(\tau) \sin \sqrt{\lambda_k}(t - \tau) \, d\tau. \tag{10.14}$$

Therefore, using the Fourier method, we obtain a solution to problem (10.1)–(10.3), (10.5) as series (10.11) with coefficients $u_k(t)$ given by formula (10.14).

Theorem 10.2. *If $\varphi \in \mathring{H}^1(Q)$, $\psi \in L_2(Q)$ and $f \in L_2(\Omega_T)$, then series* (10.11), (10.14) *converges in the space $H^1(\Omega_T)$ to a weak solution $u(x,t)$ of mixed problem* (10.1)–(10.3) *and* (10.5). *This solution satisfies the estimate*

$$\|u\|_{H^1(\Omega_T)} \leqslant C_T\big(\|\varphi\|_{\mathring{H}^1(Q)} + \|\psi\|_{L_2(Q)} + \|f\|_{L_2(\Omega_T)}\big)$$

with a positive constant C_T independent of φ, ψ and f.

Proof. First, note that

$$\sum_{k=1}^{\infty} \psi_k^2 = \|\psi\|_{L_2(Q)}^2 \tag{10.15}$$

by virtue of Parseval's identity. As we already know (see Theorem 7.3 in Section 7 of Chapter 3), the functions $e_k(x)/\sqrt{\lambda_k}$, $k = 1, 2 \ldots$, form an orthonormal basis in the space $\mathring{H}^1(Q)$ with the inner product $(u,v)_{\mathring{H}^1(Q)} = \int_Q \nabla u \nabla v\, dx$. Since

$$\varphi = \sum_{k=1}^{\infty} \varphi_k e_k = \sum_{k=1}^{\infty} \sqrt{\lambda_k}\varphi_k \frac{e_k}{\sqrt{\lambda_k}},$$

Parseval's identity in $\mathring{H}^1(Q)$ gives

$$\sum_{k=1}^{\infty} \lambda_k \varphi_k^2 = \|\varphi\|_{\mathring{H}^1(Q)}^2. \tag{10.16}$$

Further, it follows from Fubini's theorem that the function $f(\cdot,t)$ belongs to $L_2(Q)$ for almost all $t \in (0,T)$. Consequently, $f(\cdot,t)$ can be expanded into the Fourier series $f(\cdot,t) = \sum_{k=1}^{\infty} f_k(t) e_k$, and

$$\sum_{k=1}^{\infty} f_k^2(t) = \|f(\cdot,t)\|_{L_2(Q)}^2.$$

Integrating the last equality over $(0,T)$ and using Lebesgue's dominated convergence theorem and Fubini's theorem, we obtain

$$\sum_{k=1}^{\infty} \int_0^T f_k^2(t)\, dt = \|f\|_{L_2(\Omega_T)}^2. \tag{10.17}$$

The proof of the theorem consists of several steps. At the first step, we fix a number k and assume that

$$f(x,t) = f_k(t)e_k(x), \quad \varphi(x) = \varphi_k e_k(x) \quad \text{and} \quad \psi(x) = \psi_k e_k(x)$$

in (10.1), (10.2) and (10.3). We want to show that the function $u_k(t)e_k(x)$ is a weak solution to the mixed problem. This function obviously belongs to $H^1(\Omega_T)$ and has zero trace on the lateral surface Γ_T, while its trace on the lower base Q_0 of the cylinder Ω_T is equal to φ. Let us check integral identity (10.6). Integrating by parts, using relations (10.12) and Fubini's theorem, we get

$$\begin{aligned}
\int_{\Omega_T} (\nabla u \nabla v - u_t v_t)dxdt &= \int_Q dx \int_0^T (u_k(t)\nabla e_k(x)\nabla v - u_k'(t)e_k(x)v_t)dt = \\
&= \int_Q dx \int_0^T (u_k(t)\nabla e_k(x)\nabla v + u_k''(t)e_k(x)v)dt + \int_Q \psi_k e_k(x)v|_{t=0}\, dx = \\
&= \int_Q dx \int_0^T (u_k(t)\nabla e_k(x)\nabla v - \lambda_k u_k(t)e_k(x)v)dt + \\
&\quad + \int_Q \psi_k e_k(x)v|_{t=0}\, dx + \int_Q dx \int_0^T f_k(t)e_k(x)v\, dt = \\
&= \int_0^T u_k(t)\, dt \int_Q (\nabla e_k(x)\nabla v(x,t) - \lambda_k e_k(x)v(x,t))dx + \\
&\quad + \int_Q \psi v|_{t=0}\, dx + \int_{\Omega_T} fv\, dxdt
\end{aligned}$$

for $v \in \widetilde{H}^1(\Omega_T)$. Note that $v(\cdot,t)$ as a function of spatial variables x belongs to $\mathring{H}^1(Q_t)$ for almost all $t \in (0,T)$. Therefore, using the definition of a weak eigenfunction of the Dirichlet problem for the Laplace operator, we have

$$\int_Q \nabla e_k(x)\nabla v(x,t)\, dx = \lambda_k \int_Q e_k(x)v(x,t)\, dx$$

for almost all $t \in (0,T)$, and thus the desired identity is obtained.

In the second step, we consider problem (10.1), (10.2), (10.3) and (10.5), in which the functions $f(x,t)$, $\varphi(x)$ and $\psi(x)$ are replaced by partial sums of the Fourier series,

$$f(x,t) = F_N(x,t) = \sum_{k=1}^N f_k(t)e_k(x),$$

$$\varphi(x) = \Phi_N(x) = \sum_{k=1}^N \varphi_k e_k(x), \quad \psi(x) = \Psi_N(x) = \sum_{k=1}^N \psi_k e_k(x).$$

Based on the result of the first step, we see by the superposition principle that the partial sum

$$S_N(x,t) = \sum_{k=1}^{N} u_k(t)e_k(x)$$

of series (10.11) is a weak solution of the corresponding mixed problem.

In the third step, we prove that $\{S_N(x,t)\}_{N=1}^{\infty}$ is a Cauchy sequence in $H^1(\Omega_T)$. Consider the difference $S_N - S_M$, $M < N$, and note that the function $S_N(\cdot,t)$ belongs to the space $\mathring{H}^1(Q)$ for each value of t. By virtue of Corollary 6.1 on the equivalent inner product in $\mathring{H}^1(Q)$, we have

$$\|S_N - S_M\|^2_{H^1(\Omega_T)} = \int_{\Omega_T} |\nabla(S_N - S_M)|^2 dxdt + \int_{\Omega_T} (S_N - S_M)^2 dxdt + \int_{\Omega_T} ((S_N)_t - (S_M)_t)^2 dxdt \leqslant$$
$$\leqslant C_1 \int_0^T dt \int_{Q_t} |\nabla(S_N - S_M)|^2 dx + \int_0^T dt \int_{Q_t} ((S_N)_t - (S_M)_t)^2 dx.$$

Recall that

$$\int_Q e_k e_m \, dx = \delta_{km}, \quad \delta_{km} = \begin{cases} 1, & k = m, \\ 0, & k \neq m. \end{cases}$$

Thus, we arrive at the inequality

$$\|S_N - S_M\|^2_{H^1(\Omega_T)} \leqslant k_1 \sum_{k=M+1}^{N} \int_0^T (\lambda_k u_k^2(t) + (u_k'(t))^2) \, dt. \tag{10.18}$$

From the explicit expression for $u_k(t)$ (formula (10.14)), we derive

$$u_k^2(t) \leqslant k_2 \left(\varphi_k^2 + \frac{\psi_k^2}{\lambda_k} + \frac{1}{\lambda_k} \int_0^T f_k^2(\tau) \, d\tau \right),$$

$$\lambda_k u_k^2(t) \leqslant k_2 \left(\lambda_k \varphi_k^2 + \psi_k^2 + \int_0^T f_k^2(\tau) \, d\tau \right),$$

$$u_k'(t) = -\varphi_k \sqrt{\lambda_k} \sin \sqrt{\lambda_k} t + \psi_k \cos \sqrt{\lambda_k} t + \int_0^t f_k(\tau) \cos \sqrt{\lambda_k}(t - \tau) \, d\tau,$$

$$(u_k'(t))^2 \leqslant k_3 \left(\lambda_k \varphi_k^2 + \psi_k^2 + \int_0^T f_k^2(\tau) \, d\tau \right).$$

Combining these inequalities with (10.18), we end up with the following estimate:

$$\|S_N - S_M\|^2_{H^1(\Omega_T)} \leqslant k_4 \sum_{k=M+1}^{N} \left(\lambda_k \varphi_k^2 + \psi_k^2 + \int_0^T f_k^2(\tau)\, d\tau \right), \tag{10.19}$$

$$k_4 = k_1(k_2 + k_3)T.$$

Therefore, $\|S_N - S_M\|^2_{H^1(\Omega_T)} \to 0$ as $M, N \to \infty$ by (10.16), (10.15) and (10.17). Since the space $H^1(\Omega_T)$ is complete, series (10.11) converges in $H^1(\Omega_T)$. Setting $M = 0$ and passing to the limit as $N \to \infty$ in inequality (10.19), we obtain the estimate

$$\|u\|^2_{H^1(\Omega_T)} \leqslant k_4(\|\varphi\|^2_{\mathring{H}^1(Q)} + \|\psi\|^2_{L_2(Q)} + \|f\|^2_{L_2(\Omega_T)})$$

for the sum $u(x,t)$ of this series. Passing to the limit as $N \to \infty$ in the integral identity

$$\int_{\Omega_T} (\nabla S_N \nabla v - (S_N)_t v_t)\, dxdt = \int_{\Omega_T} F_N v\, dxdt + \int_Q \Psi_N v|_{t=0}\, dx$$

and the equality $S_N|_{Q_0} = \Phi_N$ (here the trace theorem is applied), we make sure that $u(x,t)$ is a weak solution of the original mixed problem. The theorem is now proven. □

Galerkin's method for wave equation

Another common method for solving mixed problems is the Galerkin method. Being an approximate method, it simultaneously provides an independent proof for the existence of a weak solution to the mixed problem.

The main advantage of the Galerkin method over the Fourier method is that it does not use the system of eigenfunctions of the corresponding elliptic problem. In addition, it can be applied to equations whose coefficients depend on both x and t. To start solving problem (10.1), (10.2), (10.3) and (10.5) by the Galerkin method, one only needs to pick a linearly independent system $v_k \in \mathring{H}^1(Q) \cap C^2(\overline{Q})$, $k = 1, 2, \ldots$, complete in $\mathring{H}^1(Q)$. The latter means that linear combinations of these functions are dense in $\mathring{H}^1(Q)$.

Suppose that $\varphi \in \mathring{H}^1(Q)$, $\psi \in L_2(Q)$ and $f \in L_2(\Omega_T)$.

The idea of the method is as follows. For any natural number m, the approximate solution $w_m(x,t)$ of problem (10.1), (10.2), (10.3) and (10.5) is sought in the form

$$w_m(x,t) = \sum_{s=1}^{m} c_s(t) v_s(x),$$

where coefficients $c_s(t)$ $(s = 1, \ldots, m)$ are functions from $H^2(0,T)$. These coefficients, in fact, depend on m, although this is not reflected in the notation. In other words, for each $t \in (0,T)$, the approximate solution belongs to the linear span V_m of the first m functions $v_1(x), \ldots, v_m(x)$. Coefficients $c_s(t)$ are chosen in such a way that the functions $w_m(x,t)$ satisfy not original relations (10.1), (10.2) and (10.3) but their $L_2(Q)$-projections onto V_m:

$$((w_m)_{tt}(\cdot,t) - \Delta w_m(\cdot,t), v_k)_{L_2(Q)} = (f(\cdot,t), v_k)_{L_2(Q)} \quad (t \in (0,T)),$$
$$(w_m(\cdot,0), v_k)_{L_2(Q)} = (\varphi, v_k)_{L_2(Q)}, \quad ((w_m)_t(\cdot,0), v_k)_{L_2(Q)} = (\psi, v_k)_{L_2(Q)}$$

for $k = 1,\dots,m$. Below we show that this system is uniquely solvable but first we write it as the initial value problem for a system of m linear second-order ordinary differential equations,

$$\sum_{s=1}^{m} (v_s, v_k)_{L_2(Q)} c_s''(t) + \sum_{s=1}^{m} (v_s, v_k)_{\mathring{H}^1(Q)} c_s(t) = (f(\cdot,t), v_k)_{L_2(Q)} \quad (t \in (0,T)), \tag{10.20}$$

$$\sum_{s=1}^{m} (v_s, v_k)_{L_2(Q)} c_s(0) = (\varphi, v_k)_{L_2(Q)}, \quad \sum_{s=1}^{m} (v_s, v_k)_{L_2(Q)} c_s'(0) = (\psi, v_k)_{L_2(Q)} \tag{10.21}$$
$$(k = 1,\dots,m).$$

Here, $(v_s, v_k)_{\mathring{H}^1(Q)} = \int_Q \nabla v_s \nabla v_k \, dx$. Problem (10.20), (10.21) represents the base of the Galerkin method from the viewpoint of its practical implementation. To justify the method, let us first make sure that this problem has a unique solution, which is a vector of functions $c_s \in H^2(0,T)$, $s = 1,\dots,m$. Let us introduce the columns

$$C(t) = (c_1(t) \dots c_m(t))^T,$$
$$F(t) = ((f(\cdot,t), v_1)_{L_2(Q)} \cdots (f(\cdot,t), v_m)_{L_2(Q)})^T,$$

where the coordinates of the vector function $F(t)$ belong to $L_2(0,T)$. By G_0 and G_1, we denote constant $m \times m$-matrices with the elements $(v_s, v_k)_{L_2(Q)}$ and $(v_s, v_k)_{\mathring{H}^1(Q)}$, respectively. Since the Gram determinant $\det G_0$ is nonzero, the initial vectors $C(0)$ and $C'(0)$ are uniquely determined from systems (10.21), and differential equations (10.20) themselves can be represented in the matrix form

$$C''(t) + G_0^{-1} G_1 C(t) = G_0^{-1} F(t). \tag{10.22}$$

It is now convenient to proceed to the $2m$-dimensional vector

$$\mathbf{C}(t) = \begin{pmatrix} C'(t) \\ C(t) \end{pmatrix}$$

to convert the system to the form

$$\mathbf{C}'(t) + \mathbf{G}\mathbf{C}(t) = \mathbf{F}(t),$$

where

$$\mathbf{G} = \begin{pmatrix} 0 & G_0^{-1} G_1 \\ -E & 0 \end{pmatrix}, \quad \mathbf{F}(t) = \begin{pmatrix} G_0^{-1} F(t) \\ 0 \end{pmatrix}.$$

Finally, move on to the equivalent matrix integral equation

$$\mathbf{C}(t) + \int_0^t \mathbf{G}\mathbf{C}(\tau)\,d\tau = \mathbf{C}(0) + \int_0^t \mathbf{F}(\tau)\,d\tau,$$

in which the vector function on the right-hand side is continuous on $[0, T]$. The existence of a unique continuous solution $\mathbf{C}(t)$ of this equation on $[0, T]$ is well known from the theory of ordinary differential equations. But from (10.22) it immediately follows that $c_s'' \in L_2(0, T)$, i. e., $c_s \in H^2(0, T)$, $s = 1, \dots, m$.

Let us now study the behavior of the constructed sequence $w_m(x, t)$, $m \to \infty$. Namely, we are going to show that this sequence converges weakly in the space $H^1(\Omega_T)$ to a weak solution $u(x, t)$ of the mixed problem under consideration.

To simplify the calculations, assume that the initial functions are equal to zero (see Exercise 45). Thus, $\varphi = \psi = 0$; hence $c_s(0) = c_s'(0) = 0$.

Multiply each of the equations

$$\int_{Q_t} (w_{mtt} - \Delta w_m) v_k\,dx = \int_{Q_t} f v_k\,dx \quad (t \in (0, T);\ k = 1, \dots, m)$$

by $c_k'(t)$, then integrate over $(0, \tau)$, and sum the obtained equalities over $k = 1, \dots, m$. We get

$$\int_{\Omega_\tau} (w_{mtt} - \Delta w_m) w_{mt}\,dxdt = \int_{\Omega_\tau} w_{mt} f\,dxdt. \tag{10.23}$$

Since

$$\begin{aligned} w_{mtt} w_{mt} &= \frac{1}{2}(w_{mt}^2)_t, \\ \Delta w_m w_{mt} &= \operatorname{div}(w_{mt} \nabla w_m) - \frac{1}{2}(|\nabla w_m|^2)_t, \end{aligned}$$

we have

$$(w_{mtt} - \Delta w_m) w_{mt} = \frac{1}{2}(w_{mt}^2 + |\nabla w_m|^2)_t - \operatorname{div}(w_{mt} \nabla w_m).$$

Hence, integrating (10.23) by parts and taking into account the initial and the boundary conditions for w_m, namely

$$w_{mt}(x, 0) = 0, \quad \nabla w_m(x, 0) = 0, \quad w_{mt}|_{\Gamma_\tau} = 0,$$

we obtain

$$\frac{1}{2} \int_{Q_\tau} (w_{mt}^2 + |\nabla w_m|^2)\,dx = \int_{\Omega_\tau} w_{mt} f\,dxdt.$$

Integrate this equality over $\tau \in (0, T)$,

$$\frac{1}{2}\int_{\Omega_T}(w_{mt}^2 + |\nabla w_m|^2)\,dxd\tau = \int_0^T d\tau \int_0^\tau dt \int_{Q_t} w_{mt} f\,dx.$$

Changing the order of integration in the iterated integral on the right-hand side and using the Cauchy–Schwarz inequality, we come to the inequality

$$\frac{1}{2}\int_{\Omega_T}(w_{mt}^2 + |\nabla w_m|^2)\,dxd\tau = \int_{\Omega_T}(T - t) w_{mt} f\,dxdt \leqslant$$
$$\leqslant T\|f\|_{L_2(\Omega_T)}\|w_m\|_{H^1(\Omega_T)}.$$

Since w_m vanishes on a part of the boundary, $w_m|_{Q_0 \cup \Gamma_T} = 0$, the integral on the left-hand side of this inequality is the square of the equivalent norm in $H^1(\Omega_T)$. Therefore, we arrive at the estimate

$$\|w_m\|_{H^1(\Omega_T)} \leqslant k_1\|f\|_{L_2(\Omega_T)}$$

with a positive constant k_1 independent of f and w_m. Therefore, the sequence w_m is bounded in $H^1(\Omega_T)$. But then it contains a subsequence weakly convergent in $H^1(\Omega_T)$ to a function $w_0 \in H^1(\Omega_T)$, $w_0|_{Q_0 \cup \Gamma_T} = 0$. Let us retain the same notation for this subsequence:

$$w_m \rightharpoonup w_0, \quad m \to \infty.$$

Let us make sure that w_0 is a weak solution of the mixed problem. In fact, it remains to check the integral identity

$$\int_{\Omega_T}(\nabla w_0 \nabla v - w_{0t} v_t)\,dxdt = \int_{\Omega_T} f v\,dxdt$$

for $v \in \widetilde{H}^1(\Omega_T)$. In order to do this, we use the formulas

$$\int_{Q_t}(w_{mtt} - \Delta w_m) v_k\,dx = \int_{Q_t} f v_k\,dx$$

valid for all $t \in (0, T)$ and $k = 1, \dots, m$. If we multiply these equalities by an arbitrary function $\theta \in C^1[0, T]$ such that $\theta(T) = 0$ and integrate over $t \in (0, T)$, then after integration by parts we obtain

$$\int_{\Omega_T}(\nabla w_m \nabla(v_k\theta) - w_{mt}(v_k\theta)_t)dxdt = \int_{\Omega_T} f v_k \theta\,dxdt.$$

For fixed k and θ, the weak convergence of w_m is sufficient to pass to the limit as $m \to \infty$ in the last relation and obtain

$$\int_{\Omega_T} (\nabla w_0 \nabla(v_k\theta) - w_{0t}(v_k\theta)_t)\, dxdt = \int_{\Omega_T} f v_k \theta\, dxdt.$$

Being valid for functions of the form $v_k\theta$, this equality is also true for linear combinations of these functions. In order to complete the reasoning, it remains to show that linear combinations of functions $v_k\theta$ for $k = 1, 2, \ldots$ and $\theta \in C^1[0, T]$, $\theta(T) = 0$ are dense in the space $\widetilde{H}^1(\Omega_T)$. Obviously, the linear span $\mathcal{M}$ of all such functions $v_k\theta$ coincides with the linear span of all functions of the form $v_k^*\theta$, where θ is the same as before and $v_1^*, v_2^*, \ldots$ is an orthonormal basis of the space $\mathring{H}^1(Q)$ with the inner product $(f, g)_{\mathring{H}^1(Q)} = \int_Q \nabla f \nabla g\, dx$, obtained from the original system by the Gram–Schmidt orthogonalization process.

Since the set of all functions from $C^2(\overline{\Omega}_T)$ that vanish on $\Gamma_T \cup Q_T$ is dense in $\widetilde{H}^1(\Omega_T)$ (Exercise 44), it suffices to check that each such function $\eta(x, t)$ can be approximated with respect to the norm

$$\|f\|_{\widetilde{H}^1(\Omega_T)} = \left(\int_{\Omega_T} (f_t^2 + |\nabla f|^2)\, dxdt \right)^{1/2}$$

by functions from $\mathcal{M}$.

For each value $t \in [0, T]$, functions $\eta(\cdot, t)$ and $\eta_t(\cdot, t)$ are elements of $\mathring{H}^1(Q)$ and are represented as Fourier series

$$\eta(x, t) = \sum_{k=1}^{\infty} \eta_k(t) v_k^*(x), \quad \eta_t(x, t) = \sum_{k=1}^{\infty} (\eta_t)_k(t) v_k^*(x).$$

Here,

$$\eta_k(t) = \int_Q \nabla\eta(x, t) \nabla v_k^*(x)\, dx,$$

$$(\eta_t)_k(t) = \int_Q \nabla\eta_t(x, t)\nabla v_k^*(x)\, dx = \frac{d}{dt}\left(\int_Q \nabla\eta(x, t)\nabla v_k^*(x)\, dx \right) = \eta_k'(t).$$

Parseval's identity implies

$$\sum_{k=1}^{\infty} (\eta_k^2(t) + \eta_k'^2(t)) = \int_Q \left(|\nabla\eta(x, t)|^2 + |\nabla\eta_t(x, t)|^2 \right) dx, \quad t \in [0, T].$$

Lebesgue's dominated convergence theorem allows us to integrate the series on the left-hand side term by term:

$$\sum_{k=1}^{\infty} \int_0^T (\eta_k^2(t) + \eta_k'^2(t))\, dt < \infty.$$

Let us denote $\eta_N = \sum_{k=1}^N \eta_k v_k^*$. By virtue of the Friedrichs inequality, there exists a constant $c > 0$ depending only on Q and such that

$$\|\eta_t - (\eta_N)_t\|_{L_2(Q_t)} \leqslant c\|\eta_t - (\eta_N)_t\|_{\mathring{H}^1(Q_t)}.$$

Therefore,

$$\begin{aligned}\|\eta - \eta_N\|^2_{\tilde{H}^1(\Omega_T)} &= \int_0^T (\|\eta_t - (\eta_N)_t\|^2_{L_2(Q_t)} + \|\eta - \eta_N\|^2_{\mathring{H}^1(Q_t)})\, dt \leqslant \\ &\leqslant \int_0^T (c^2\|\eta_t - (\eta_N)_t\|^2_{\mathring{H}^1(Q_t)} + \|\eta - \eta_N\|^2_{\mathring{H}^1(Q_t)})\, dt = \\ &= \sum_{k=N+1}^{\infty} \int_0^T (c^2\eta_k'^2(t) + \eta_k^2(t))dt \to 0, \quad N \to \infty.\end{aligned}$$

In fact, we have shown that every weakly convergent subsequence of the sequence w_m converges (weakly) to a weak solution w_0 of the mixed problem. But in view of uniqueness of weak solution shown above, it turns out that the entire sequence converges weakly to the weak solution. Indeed, suppose on the contrary that such convergence does not take place. Then there exists a continuous linear functional Λ on the space $H^1(\Omega_T)$ such that $\Lambda(w_m - w_0) \not\to 0$ as $m \to \infty$, i. e., $|\Lambda(w_{m_p} - w_0)| \geqslant \varepsilon$ for some number $\varepsilon > 0$ and some subsequence w_{m_p} of the sequence w_m. But we already know that w_{m_p} itself contains a subsequence that weakly converges to w_0, and that is a contradiction.

Thus we arrive at the following result.

Theorem 10.3. *Let $\varphi \in \mathring{H}^1(Q)$, $\psi \in L_2(Q)$ and $f \in L_2(\Omega_T)$. Then for an arbitrary linearly independent system $v_k \in \mathring{H}^1(Q) \cap C^2(\overline{Q})$, $k = 1, 2, \ldots$, complete in $\mathring{H}^1(Q)$, a sequence $w_m(x,t) = \sum_{s=1}^m c_s(t)v_s(x)$ of approximations of the Galerkin method, where $c_s(t)$ are determined for each m in accordance with* (10.20) *and* (10.21), *weakly converges in the space $H^1(\Omega_T)$ to a weak solution of problem* (10.1), (10.2), (10.3) *and* (10.5).

Remark 10.1. The choice of a system of basis functions $v_k(x)$ affects the nature of convergence for the sequence of the Galerkin approximations. For example, if we take the eigenfunctions of the corresponding boundary-value problem for the Laplace operator as v_k, then the coefficients $c_k(t)$ coincide with the coefficients $u_k(t)$ of the Fourier method, and the sequence of approximations of the Galerkin method, being a sequence of partial sums of the Fourier series, converges to the weak solution of the mixed problem already in the norm of the space $H^1(\Omega_T)$.

Remark 10.2. Taking into account all kinds of generalizations and modifications united by a common idea that finite-dimensional approximations are found from the condition of "orthogonality" of residuals to finite-dimensional subspaces, the Galerkin method is a powerful tool for studying the solvability of a wide range of problems, including nonlinear equations and operator equations [6, Chapter 3, Section 3].

11 Mixed problems for the heat equation

When studying mixed problems for the heat equation, we generally adhere to the same scheme as for the wave equation. However, in contrast to the latter, the heat equation is an equation of second order with respect to spatial variables and only first order with respect to time. It is natural to expect that the solutions of the heat equation differ in smoothness with respect to x and t. Therefore, in this section we introduce *anisotropic Sobolev spaces.*

Anisotropic Sobolev spaces

Let Q again be a bounded domain in $\mathbb{R}^n$, $\partial Q \in C^1$, Ω_T be the cylinder $Q \times (0,T)$, and functions $u = u(x,t)$ be defined in Ω_T. By definition, the anisotropic Sobolev space $H^{1,0}(\Omega_T)$ consists of all real-valued functions $u \in L_2(\Omega_T)$ possessing weak derivatives $u_{x_i} \in L_2(\Omega_T)$ in spatial variables, i. e.,

$$H^{1,0}(\Omega_T) = \{u \in L_2(\Omega_T) : u_{x_i} \in L_2(\Omega_T),\ i = 1, \dots, n\}.$$

Note that the existence of weak derivative $u_t \in L_2(\Omega_T)$ with respect to the variable t is not required. The space $H^{1,0}(\Omega_T)$ is equipped with the inner product

$$(u, v)_{H^{1,0}(\Omega_T)} = \int_{\Omega_T} (uv + \nabla u \nabla v)\, dxdt.$$

Remark 11.1. The space $H^{1,0}(\Omega_T)$ can be identified with the Lebesgue space $L_2((0,T); H^1(Q))$ consisting of all $H^1(Q)$-valued functions on the interval $(0,T)$, with the inner product

$$(u, v)_{L_2((0,T);H^1(Q))} = \int_0^T (u(\cdot,t), v(\cdot,t))_{H^1(Q)} dt = (u, v)_{H^{1,0}(\Omega_T)}.$$

By Fubini's theorem,

$$(u,v)_{H^{1,0}(\Omega_T)} = \int_0^T dt \int_{Q_t} (uv + \nabla u \nabla v)\,dx \quad (u,v \in H^{1,0}(\Omega_T)). \tag{11.1}$$

In view of relation (11.1), the validity of Remark 11.1 follows from the following two statements:

(a) if $u \in H^{1,0}(\Omega_T)$, then the function $u_{x_i}(\cdot,t)$ is the weak derivative of the function $u(\cdot,t)$ in the domain Q for almost all $t \in (0,T)$;

(b) if $u \in L_2((0,T);H^1(Q))$ and $u_{x_i}(\cdot,t)$ is the weak derivative of the function $u(\cdot,t)$ in the domain Q for almost all $t \in (0,T)$, then there exists the weak derivative $u_{x_i} \in L_2(\Omega_T)$ equal to $u_{x_i}(\cdot,t)$ for almost all $t \in (0,T)$.

We restrict ourselves to proving statement (a). Let $u_{x_i} \in L_2(\Omega_T)$ be the weak derivative of a function $u \in H^{1,0}(\Omega_T)$. Let us choose a test function in the form of $\eta(t)\varphi(x)$ with arbitrary $\eta \in C_0^\infty(0,T)$ and $\varphi \in C_0^\infty(Q)$, and substitute it for v in the integral identity

$$\int_{\Omega_T} uv_{x_i}\,dxdt = -\int_{\Omega_T} u_{x_i} v\,dxdt.$$

Then we have

$$\int_0^T \eta(t)\,dt \int_{Q_t} (u\varphi_{x_i} + u_{x_i}\varphi)\,dx = 0. \tag{11.2}$$

Taking into account that $u(\cdot,t)$ and $u_{x_i}(\cdot,t)$ as functions of spatial variables x belong to $L_2(Q)$ for almost all $t \in (0,T)$, and the modulus of the internal integral $I(t) = \int_{Q_t}(u\varphi_{x_i} + u_{x_i}\varphi)\,dx$ is estimated by the expression $k_1(\|u(\cdot,t)\|_{L_2(Q)} + \|u_{x_i}(\cdot,t)\|_{L_2(Q)})$, we see that $I \in L_2(0,T)$. Since $C_0^\infty(0,T)$ is dense in $L_2(0,T)$, it follows from (11.2) that I vanishes for almost all $t \in (0,T)$; hence the equality

$$\int_Q u(x,t)\varphi_{x_i}(x)\,dx = -\int_Q u_{x_i}(x,t)\varphi(x)\,dx$$

holds on the entire interval $(0,T)$, with the possible exception of a subset of zero Lebesgue measure. Since the union of a countable family of zero measure sets also has measure zero, the last equality holds a. e. on $(0,T)$ when the function $\varphi \in C_0^\infty(Q)$ runs over a countable system complete in $\mathring{H}^1(Q)$. Consequently, it holds for all $\varphi \in C_0^\infty(Q)$ a. e. on $(0,T)$. We have thereby shown that the function $u_{x_i}(\cdot,t)$ is the weak derivative of the function $u(\cdot,t)$ for almost all $t \in (0,T)$.

The results of the second chapter of the textbook are easy to adapt to the case of the anisotropic space $H^{1,0}(\Omega_T)$.

Theorem 11.1. *The space $H^{1,0}(\Omega_T)$ is complete with respect to the norm*

$$\|u\|_{H^{1,0}(\Omega_T)} = \left\{ \int\limits_{\Omega_T} (u^2 + |\nabla u|^2)\, dxdt \right\}^{1/2},$$

i. e., is a Hilbert space.

Theorem 11.2. *Let $\partial Q \in C^1$ and Q' be a bounded domain in $\mathbb{R}^n$ such that $\overline{Q} \subset Q'$. If Ω'_{t_0,t_1} is the larger cylinder $Q' \times (t_0, t_1)$, $t_0 < 0$ and $t_1 > T$, then any function $f \in H^{1,0}(\Omega_T)$ can be extended to Ω'_{t_0,t_1} by a compactly supported function $F \in H^{1,0}(\Omega'_{t_0,t_1})$ such that*

$$\|F\|_{H^{1,0}(\Omega'_{t_0,t_1})} \leqslant c_1 \|f\|_{H^{1,0}(\Omega_T)}.$$

Theorem 11.3. *The set $C^\infty(\overline{\Omega}_T)$ is dense in the space $H^{1,0}(\Omega_T)$.*

Unlike the isotropic Sobolev space $H^1(Q)$, the trace of a function from the space $H^{1,0}(\Omega_T)$ cannot be defined on the entire boundary $\partial\Omega_T$ in the sense of Section 6. However, as further considerations show, it is possible to introduce the concept of the trace of a function from $H^{1,0}(\Omega_T)$ on the lateral surface Γ_T of the cylinder.

By virtue of Theorem 6.6, the estimate

$$\|f\|_{L_2(\partial Q_t)} \leqslant k_1 \|f\|_{H^1(Q_t)} \tag{11.3}$$

holds for an arbitrary function $f \in C^1(\overline{\Omega}_T)$. Since all sections Q_t are identical, the constant $k_1 > 0$ does not depend on either f or $t \in (0, T)$. Integrating over $(0, T)$ the squared inequality (11.3) and taking into account the definition of the norm in $H^{1,0}(\Omega_T)$, we obtain

$$\|f\|_{L_2(\Gamma_T)} \leqslant k_1 \|f\|_{H^{1,0}(\Omega_T)}.$$

We then approximate an arbitrary function $f \in H^{1,0}(\Omega_T)$ in $H^{1,0}(\Omega_T)$ by a sequence of smooth functions $f_m \in C^1(\overline{\Omega}_T)$ such that

$$\|f_m\|_{L_2(\Gamma_T)} \leqslant k_1 \|f_m\|_{H^{1,0}(\Omega_T)}, \quad \|f_m - f_p\|_{L_2(\Gamma_T)} \leqslant k_1 \|f_m - f_p\|_{H^{1,0}(\Omega_T)}.$$

The restrictions $f_m|_{\Gamma_T}$ thereby form a Cauchy sequence in $L_2(\Gamma_T)$. By definition, the limit of this sequence is the *trace* $f|_{\Gamma_T}$ of the function $f \in H^{1,0}(\Omega_T)$ on the lateral surface Γ_T of the cylinder. The trace does not depend on the choice of an approximating sequence and satisfies the estimate

$$\|f|_{\Gamma_T}\|_{L_2(\Gamma_T)} \leqslant C \|f\|_{H^{1,0}(\Omega_T)}. \tag{11.4}$$

Thus, we have proven the following statement.

Theorem 11.4. *Let $Q \subset \mathbb{R}^n$ be a bounded domain with boundary of class C^1. Then for any function $f \in H^{1,0}(\Omega_T)$, its trace $f|_{\Gamma_T} \in L_2(\Gamma_T)$ is defined. The trace operator is bounded from $H^{1,0}(\Omega_T)$ to $L_2(\Gamma_T)$.*

For functions from $H^{1,0}(\Omega_T)$ with zero trace on Γ_T, the equivalent norm

$$\|f\|^2_{H^{1,0}(\Omega_T)} = \int_0^T \|f\|^2_{\mathring{H}^1(Q_t)} dt = \int_0^T \||\nabla f|\|^2_{L_2(Q_t)} dt = \int_{\Omega_T} |\nabla f|^2 dxdt$$

in $H^{1,0}(\Omega_T)$ is used in this section.

Theorem 11.5. *The integration by parts formula*

$$\int_{\Omega_T} f_{x_i} g\, dxdt = \int_{\Gamma_T} f|_{\Gamma_T} g|_{\Gamma_T} \nu_i\, dSdt - \int_{\Omega_T} f g_{x_i}\, dxdt$$

is valid for all functions $f, g \in H^{1,0}(\Omega_T)$, where $\nu = (\nu_1, \dots \nu_n, 0)$ is the unit vector of the outer normal to the lateral surface Γ_T.

Uniqueness of the weak solution

In the case of the first mixed problem, a solution to the heat equation

$$u_t - \Delta u = f(x,t) \quad ((x,t) \in \Omega_T) \tag{11.5}$$

is sought in a cylinder and satisfies the initial condition

$$u|_{t=0} = \varphi(x) \quad (x \in Q) \tag{11.6}$$

on the lower base of the cylinder and the boundary condition

$$u|_{\Gamma_T} = 0 \tag{11.7}$$

on its lateral surface.

A function $u \in C^{2,1}(\Omega_T) \cap C(\Omega_T \cup \Gamma_T \cup \overline{Q}_0)$ is called the classical solution to problem (11.5)–(11.7) if it satisfies equation (11.5) on Ω_T, initial condition (11.6) on Q_0 and boundary condition (11.7) on Γ_T. Here, $C^{2,1}(\Omega_T)$ is the set of all functions continuous in Ω_T whose derivatives u_{x_i}, $u_{x_i x_j}$ $(i,j = 1, \dots, n)$, and u_t are also continuous in Ω_T.

Let a function u belong to the space $C^{2,1}(\overline{\Omega}_T)$ and be a classical solution to problem (11.5)–(11.7). Multiply equation (11.5) by an arbitrary function $v \in \widetilde{H}^1(\Omega_T)$ and integrate over Ω_T. Integrating by parts in the integral $\int_{\Omega_T} (u_t v - \Delta u v)\, dxdt$, we come to the relation

$$\int_{\Omega_T} (\nabla u \nabla v - u v_t)\, dxdt = \int_{\Omega_T} f v\, dxdt + \int_Q \varphi v|_{t=0}\, dx. \tag{11.8}$$

Definition 11.1. Suppose that $f \in L_2(\Omega_T)$ and $\varphi \in L_2(Q)$. A function $u \in H^{1,0}(\Omega_T)$ is called a *weak solution* of problem (11.5), (11.6) and (11.7) if it satisfies integral identity (11.8) for all functions $v \in \widetilde{H}^1(\Omega_T)$ and has zero trace on Γ_T.

Remark 11.2. In the above definition, the weak solution belongs to the anisotropic Sobolev space, while the test function v is an arbitrary element of the ordinary, isotropic, Sobolev space. The space of test functions for the heat equation is the same as for the wave equation.

Remark 11.3. A weak solution satisfies the boundary condition in the sense of trace, and the initial value function is present only in the integral identity.

Theorem 11.6. *Problem* (11.5)–(11.7) *has at most one weak solution.*

Proof. It suffices to show that the only weak solution u of the homogeneous problem for $f = 0$ and $\varphi = 0$ is the trivial solution. For this purpose, we select the test function as follows:

$$v(x,t) = \int_t^T u(x,\tau)\, d\tau.$$

As a result of integration, the function v becomes smooth with respect to t, thus $v \in \widetilde{H}^1(\Omega_T)$, $v_t = -u$, and $v_{x_i} = \int_t^T u_{x_i}(x,\tau)\, d\tau$. The integral identity for the function u takes the form

$$\int_Q dx \int_0^T \nabla u(x,t)\, dt \int_t^T \nabla u(x,\tau)\, d\tau + \int_{\Omega_T} u^2\, dxdt = 0. \tag{11.9}$$

Changing the order of integration in the iterated integral

$$I = \int_0^T \nabla u(x,t)\, dt \int_t^T \nabla u(x,\tau)\, d\tau,$$

we get

$$I = \int_0^T \nabla u(x,\tau)\, d\tau \int_0^\tau \nabla u(x,t)\, dt = \int_0^T \nabla u(x,\tau)\, d\tau \int_0^T \nabla u(x,t)\, dt -$$

$$- \int_0^T \nabla u(x,\tau)\, d\tau \int_\tau^T \nabla u(x,t)\, dt = \left| \int_0^T \nabla u(x,t)\, dt \right|^2 - I,$$

whence it follows that $I = \frac{1}{2}|\int_0^T \nabla u(x,t)\,dt|^2 \geqslant 0$, and hence the first term on the left-hand side of (11.9) is nonnegative. It follows now from (11.9) that u is zero a. e. in Ω_T. □

Fourier method for heat equation

Regarding the algorithm, the Fourier method for solving the heat equation is exactly the same as the corresponding method used for solving the wave equation. The solution to problem (11.5)–(11.7) is sought in the form of a series

$$u(x,t) = \sum_{k=0}^{\infty} u_k(t)e_k(x), \tag{11.10}$$

where $e_k(x)$ are the weak eigenfunctions of problem (10.9), (10.10), which form an orthonormal basis in the space $L_2(Q)$. Formal substitution of this series in equation (11.5) and condition (11.6) leads to the following relations:

$$\sum_{k=1}^{\infty} u_k'(t)e_k(x) + \sum_{k=1}^{\infty} \lambda_k u_k(t)e_k(x) = \sum_{k=1}^{\infty} f_k(t)e_k(x),$$
$$\sum_{k=0}^{\infty} u_k(0)e_k(x) = \sum_{k=1}^{\infty} \varphi_k e_k(x).$$

From here, we obtain the family

$$u_k'(t) + \lambda_k u_k(t) = f_k(t), \quad u_k(0) = \varphi_k \quad (k = 1,2,\dots) \tag{11.11}$$

of initial value problems for ordinary differential equations on the interval $(0,T)$, where, as before, the numbers λ_k are the eigenvalues of problem (10.9), (10.10) and $f_k(t)$ and φ_k are the Fourier coefficients of functions $f(x,t)$ and $\varphi(x)$ with respect to the orthonormal basis $\{e_k(x)\}_{k=1}^{\infty}$ in $L_2(Q)$. Solving these problems, we obtain

$$u_k(t) = \varphi_k e^{-\lambda_k t} + \int_0^t e^{-\lambda_k(t-\tau)} f_k(\tau)\,d\tau. \tag{11.12}$$

Let us write down an estimate for $u_k(t)$ needed to prove the theorem below. By virtue of the Cauchy–Schwarz inequality, we have

$$u_k^2(t) \leqslant 2\varphi_k^2 e^{-2\lambda_k t} + 2\int_0^T f_k^2(\tau)\,d\tau \int_0^t e^{-2\lambda_k(t-\tau)}\,d\tau \leqslant$$
$$\leqslant 2\varphi_k^2 e^{-2\lambda_k t} + \frac{1}{\lambda_k}\int_0^T f_k^2(\tau)\,d\tau.$$

Therefore,

$$\int_0^T u_k^2(t)\,dt \leqslant \frac{1}{\lambda_k}\varphi_k^2 + \frac{T}{\lambda_k}\int_0^T f_k^2(\tau)\,d\tau. \tag{11.13}$$

Theorem 11.7. *If $\varphi \in L_2(Q)$ and $f \in L_2(\Omega_T)$, then series (11.10), (11.12) converges in the space $H^{1,0}(\Omega_T)$ to a weak solution $u(x,t)$ of mixed problem (11.5)–(11.7), and the estimate*

$$\|u\|_{H^{1,0}(\Omega_T)} \leqslant C_T\big(\|\varphi\|_{L_2(Q)} + \|f\|_{L_2(\Omega_T)}\big)$$

holds with a constant $C_T > 0$ independent of φ and f.

Remark 11.4. Unlike the wave equation, the initial value function φ is now an arbitrary function from $L_2(Q)$.

Proof. The proof follows the same scheme as for the wave equation. We start with the case where the Fourier series for f and φ consist of only one term, namely $f(x,t) = f_k(t)e_k(x)$, $\varphi(x) = \varphi_k e_k(x)$. We want to show that the function $u(x,t) = u_k(t)e_k(x)$ is a weak solution to the mixed problem, where $u_k(t)$ has form (11.12). Let us check the integral identity. Integrating by parts and substituting $-\lambda_k u_k(t) + f_k(t)$ for $u_k'(t)$, and φ_k for $u_k(0)$, we bring the integral on the left-hand side of (11.8) to the form

$$\begin{aligned}
&\int_{\Omega_T}(-u_k e_k v_t + u_k\nabla e_k\nabla v)\,dxdt = \\
&\quad = \int_{\Omega_T}(u_k' e_k v + u_k\nabla e_k\nabla v)\,dxdt + \int_Q u_k(0)e_k v\big|_{t=0}dx = \\
&\quad = \int_0^T u_k(t)\,dt\int_Q(-\lambda_k e_k(x)v(x,t) + \nabla e_k(x)\nabla v(x,t))\,dx + \\
&\quad + \int_{\Omega_T} f_k e_k v\,dxdt + \int_Q \varphi_k e_k v|_{t=0}dx.
\end{aligned}$$

Since $v \in \widetilde{H}^1(\Omega_T)$, we have $v(\cdot,t) \in \mathring{H}^1(Q)$ for almost all $t \in (0,T)$. It remains to recall the definition of a weak eigenfunction, according to which the inner integral in the first term of the last expression vanishes for almost all $t \in (0,T)$.

According to the superposition principle, partial sums S_N of series (11.10) are weak solutions of problem (11.5)–(11.7) when f and φ are replaced by partial sums F_N and Φ_N of the corresponding Fourier series.

Let us prove the convergence of series (11.10) in the space $H^{1,0}(\Omega_T)$. Moving on to the iterated integral, using the Friedrichs inequality in $\mathring{H}^1(Q)$ and the basic properties of weak eigenfunctions e_k, we can write $(N > M)$

$$\|S_N - S_M\|^2_{H^{1,0}(\Omega_T)} = \int\limits_{\Omega_T} \left((S_N - S_M)^2 + |\nabla(S_N - S_M)|^2\right) dxdt =$$

$$= \int\limits_0^T \|S_N - S_M\|^2_{\mathring{H}^1(Q_t)} dt \leqslant k_1 \int\limits_0^T dt \int\limits_{Q_t} |\nabla(S_N - S_M)|^2 \, dx =$$

$$= k_1 \int\limits_0^T dt \int\limits_Q \left|\nabla\left(\sum_{k=M+1}^N u_k(t) e_k(x)\right)\right|^2 dx = k_1 \sum_{k=M+1}^N \lambda_k \int\limits_0^T u_k^2(t) dt.$$

Based on (11.13), we obtain the inequality

$$\|S_N - S_M\|^2_{H^{1,0}(\Omega_T)} \leqslant k_1 \sum_{k=M+1}^N \left(\varphi_k^2 + T \int\limits_0^T f_k^2(\tau) \, d\tau\right). \tag{11.14}$$

It follows from (11.14), (10.17) and the Parseval identity

$$\|\varphi\|^2_{L_2(Q)} = \sum_{k=1}^\infty \varphi_k^2$$

that

$$\|S_N - S_M\|^2_{H^{1,0}(\Omega_T)} \to 0, \quad M, N \to \infty.$$

Since the space $H^{1,0}(\Omega_T)$ is complete, series (11.10) converges in $H^{1,0}(\Omega_T)$ to a function $u(x,t)$. The estimate

$$\|u\|^2_{H^{1,0}(\Omega_T)} \leqslant C_1\left(\|\varphi\|^2_{L_2(Q)} + T\|f\|^2_{L_2(\Omega_T)}\right)$$

follows from (11.14), (10.17) and the Parseval identity for φ in the limit as $N \to \infty$, $M = 0$. Passing to the limit as $N \to \infty$ in the integral identity

$$\int\limits_{\Omega_T} (\nabla S_N \nabla v - S_N v_t) \, dxdt = \int\limits_{\Omega_T} F_N v \, dxdt + \int\limits_Q \Phi_N v|_{t=0} \, dx,$$

we ascertain that $u(x,t)$ is a weak solution to problem (11.5)–(11.7). □

Galerkin's method for the heat equation

Similar to the case of the wave equation, choose an arbitrary linearly independent system of functions $v_k \in \mathring{H}^1(Q) \cap C^2(\overline{Q})$, $k = 1, 2, \ldots$, complete in $\mathring{H}^1(Q)$. As before, V_m denotes the linear span of the first m elements of this system. For each $m = 1, 2, \ldots$, we solve the problem obtained by orthogonal projection of the original mixed problem onto V_m in the space $L_2(Q)$. In other words, we are going to find a function of the form $w_m(x,t) = \sum_{s=1}^m c_s(t)v_s(x)$, $c_s \in H^1(0,T)$, by equating the projections onto V_m of the expressions on the right- and left-hand sides of relations (11.5) and (11.6),

$$(w_{mt}(\cdot,t) - \Delta w_m(\cdot,t), v_k)_{L_2(Q)} = (f(\cdot,t), v_k)_{L_2(Q)} \quad (t \in (0,T)),$$
$$(w_m(\cdot,0), v_k)_{L_2(Q)} = (\varphi, v_k)_{L_2(Q)}, \quad k = 1,\dots,m.$$

The result at each step is the following initial value problem for a linear system of m ordinary differential equations:

$$\sum_{s=1}^m (v_s, v_k)_{L_2(Q)} c_s'(t) + \sum_{s=1}^m (v_s, v_k)_{\mathring{H}^1(Q)} c_s(t) =$$
$$= (f(\cdot,t), v_k)_{L_2(Q)} \quad (t \in (0,T)), \tag{11.15}$$
$$\sum_{s=1}^m (v_s, v_k)_{L_2(Q)} c_s(0) = (\varphi, v_k)_{L_2(Q)} \tag{11.16}$$
$$(k = 1,\dots,m).$$

Here, $(v_s, v_k)_{\mathring{H}^1(Q)} = \int_Q \nabla v_s \nabla v_k \, dx$. Realizing that the value of $c_s(t)$ may depend on m (namely the coefficient $c_1(t)$ obtained at the first step differs from the function $c_1(t)$ obtained at the second step), we omit the index m for a more concise notation.

Using the former matrix notation (see Section 10), we write

$$G_0 C'(t) + G_1 C(t) = F(t) \quad (t \in (0,T)),$$
$$G_0 C(0) = \Phi := \big((\varphi, v_1)_{L_2(Q)}, \dots (\varphi, v_m)_{L_2(Q)}\big)^T$$

or

$$C'(t) + G_0^{-1} G_1 C(t) = G_0^{-1} F(t) \quad (t \in (0,T)),$$
$$C(0) = G_0^{-1}\Phi.$$

Analogously to (10.20) and (10.21), it is easy to show that the this problem has a unique solution $C(t)$, and $c_s \in H^1(0,T)$ $(s = 1,\dots,m)$.

Then we prove the weak convergence in the space $H^{1,0}(\Omega_T)$ of the constructed sequence $w_m(x,t)$ of approximations to a weak solution of the mixed problem. Following a familiar scheme, we first show that this sequence is bounded in $H^{1,0}(\Omega_T)$. For now, we do not assume that the function φ is zero in order to demonstrate how the initial value function is taken into account in the reasoning.

For a fixed m, we multiply the equations

$$\int_{Q_t} (w_{mt} - \Delta w_m) v_k \, dx = \int_{Q_t} f v_k \, dx \quad (k \leqslant m;\ t \in (0,T)) \tag{11.17}$$

by $c_k(t)$, integrate over t, and sum over $k = 1, \dots, m$ to obtain

$$\int_{\Omega_T} (w_{mt} - \Delta w_m) w_m \, dxdt = \int_{\Omega_T} f w_m \, dxdt.$$

We integrate by parts in the integral on the left-hand side. Taking into account the obvious relations $(w_m)_t w_m = \frac{1}{2}(w_m^2)_t$ and $w_m|_{\Gamma_T} = 0$, we arrive at the equality

$$\int_{\Omega_T} \left(\frac{1}{2}(w_m^2)_t + |\nabla w_m|^2 \right) dxdt = \int_{\Omega_T} f w_m \, dxdt$$

or

$$\frac{1}{2} \int_{Q_T} w_m^2 dx + \int_{\Omega_T} |\nabla \bar{w}_m|^2 dxdt = \int_{\Omega_T} f w_m \, dxdt + \frac{1}{2} \int_{Q_0} w_m^2 dx. \tag{11.18}$$

Omitting the first term on the left-hand side of the last relation and using an elementary algebraic inequality $2ab \leqslant \varepsilon^{-1} a^2 + \varepsilon b^2$, we obtain

$$\int_{\Omega_T} |\nabla w_m|^2 dxdt \leqslant \frac{1}{2\varepsilon} \|f\|^2_{L_2(\Omega_T)} + \frac{\varepsilon}{2} \|w_m\|^2_{L_2(\Omega_T)} + \frac{1}{2} \|w_m(\cdot, 0)\|^2_{L_2(Q)}. \tag{11.19}$$

Integrating the Friedrichs inequality,

$$\int_{Q_t} w_m^2 dx \leqslant d^2 \int_{Q_t} |\nabla w_m|^2 dx,$$

($d = \operatorname{diam} Q$) over $t \in (0, T)$, we get

$$\|w_m\|^2_{L_2(\Omega_T)} \leqslant d^2 \int_{\Omega_T} |\nabla w_m|^2 \, dxdt. \tag{11.20}$$

It follows from the construction of w_m that the initial value $w_m(x, 0)$ is the orthogonal projection in $L_2(Q)$ of the function φ onto the finite-dimensional subspace V_m. Hence,

$$\|w_m(\cdot, 0)\|_{L_2(Q)} \leqslant \|\varphi\|_{L_2(Q)}. \tag{11.21}$$

Combining (11.19) with (11.20) and (11.21), we can write

$$(1 - d^2\varepsilon/2) \int_{\Omega_T} |\nabla w_m|^2 \, dxdt \leqslant \frac{1}{2\varepsilon} \|f\|^2_{L_2(\Omega_T)} + \frac{1}{2} \|\varphi\|^2_{L_2(Q)}.$$

Setting $\varepsilon = d^{-2}$, we arrive at the estimate

$$\int_{\Omega_T} |\nabla w_m|^2 \, dxdt \leqslant d^2 \|f\|^2_{L_2(\Omega_T)} + \|\varphi\|^2_{L_2(Q)}.$$

In view of (11.20), the sequence w_m is bounded in $H^{1,0}(\Omega_T)$ and contains a subsequence (for which the same notation w_m is used) weakly convergent in the space $H^{1,0}(\Omega_T)$ to a function $w_0 \in H^{1,0}(\Omega_T)$.

Multiplying equations (11.17) by an arbitrary function $\theta \in C^1[0,T]$ such that $\theta(T) = 0$, then integrating over $t \in (0,T)$ and applying the integration by parts formula, we arrive at the equalities

$$\int_{\Omega_T} (\nabla w_m \nabla(v_k\theta) - w_m(v_k\theta)_t) \, dxdt = \int_{\Omega_T} f v_k \theta \, dxdt + \int_Q w_m(x,0) v_k(x)\theta(0) \, dx$$

for $m \geqslant k$. Being complete in $\mathring{H}^1(Q)$, the system $\{v_k, k = 1, 2, \dots\}$, is obviously complete in $L_2(Q)$. This means that the projections $w_m(x,0)$ of the function φ onto V_m approximate φ in $L_2(Q)$ as $m \to \infty$. Consequently, one can pass to the limit as $m \to \infty$ in each term of the last relation and obtain

$$\int_{\Omega_T} (\nabla w_0 \nabla(v_k\theta) - w_0(v_k\theta)_t) \, dxdt = \int_{\Omega_T} f v_k \theta \, dxdt + \int_Q \varphi(x) v_k(x)\theta(0) \, dx. \tag{11.22}$$

We then move from individual functions $v_k\theta$ to their linear combinations in this identity. We already know from the previous section that these linear combinations approximate any function $v \in \widetilde{H}^1(\Omega_T)$ in the $H^1(\Omega_T)$ norm. By the trace theorem, $v|_{Q_0}$ is then approximated by the traces of the corresponding linear combinations on the lower base Q_0 of the cylinder. Thus, identity (11.22) holds for all functions $v \in \widetilde{H}^1(\Omega_T)$,

$$\int_{\Omega_T} (\nabla w_0 \nabla v - w_0 v_t) \, dxdt = \int_{\Omega_T} f v \, dxdt + \int_Q \varphi v|_{t=0} \, dx.$$

Therefore, w_0 is a weak solution of the mixed problem and, similar to the previous section, it is proved that the sequence w_m weakly converges in the space $H^{1,0}(\Omega_T)$ to this weak solution.

Theorem 11.8. *For an arbitrarily chosen linearly independent system $v_k \in \mathring{H}^1(Q) \cap C^2(\overline{Q})$, $k = 1, 2, \dots$, complete in $\mathring{H}^1(Q)$, the sequence $w_m(x,t) = \sum_{s=1}^m c_s(t) v_s(x)$ of approximations of the Galerkin method, where the coefficients $c_s(t)$ are determined for all m in accordance with* (11.15) *and* (11.16), *weakly converges in the space $H^{1,0}(\Omega_T)$ to a weak solution of problem* (11.5)–(11.7).

Remark 11.5. At the end of this section, we present one important consequence of the Galerkin method. It makes no sense to talk about the values of a function u in a fixed section Q_t of the cylinder if we are talking about an arbitrary function $u \in H^{1,0}(\Omega_T)$.

However, the restriction $u|_{Q_t}$ can be defined (in a weaker sense than the trace of a function from $H^1(\Omega_T)$) for a weak solution of the mixed problem.

Let again $w_m(x,t)$, $m = 1, 2, \ldots$, denote the Galerkin approximations of a weak solution $u(x,t)$, $u \in H^{1,0}(\Omega_T)$, of problem (11.5)–(11.7), corresponding to a chosen system $\{v_k(x),\ k = 1, 2 \ldots\}$. In fact, a stronger estimate follows from (11.18)–(11.21):

$$\|w_m(\cdot, T)\|^2_{L_2(Q)} + \int_{\Omega_T} |\nabla w_m|^2\, dxdt \leqslant d^2 \|f\|^2_{L_2(\Omega_T)} + \|\varphi\|^2_{L_2(Q)}.$$

This estimate means that the sequence of pairs $\{w_m, w_m(\cdot, T)\}$ is bounded in the Cartesian product $H^{1,0}(\Omega_T) \times L_2(Q)$. Moving on to the subsequence, we assume that $\{w_m, w_m(\cdot, T)\}$ weakly converges in $H^{1,0}(\Omega_T) \times L_2(Q)$. The first coordinate $u \in H^{1,0}(\Omega_T)$ of the limit vector-function $\{u, u|_{Q_T}\}$ is, as we already know, a weak solution of the mixed problem, while $u|_{Q_T}$ is an element of $L_2(Q)$. Let us make sure that this element is uniquely determined by the problem.

Let us turn again to relations (11.17). Without multiplication by the function $\theta(t)$ this time, we integrate over $t \in (0, T)$ and, after integrating by parts, get

$$\int_{\Omega_T} \nabla w_m \nabla v_k\, dxdt + \int_Q w_m(x, T) v_k(x)\, dx = \int_{\Omega_T} f v_k\, dxdt + \int_Q w_m(x, 0) v_k(x)\, dx$$

for $m \geqslant k$. Since the sequence $\{w_m, w_m(\cdot, T)\}$ weakly converges in $H^{1,0}(\Omega_T) \times L_2(Q)$, we can pass to the limit as $m \to \infty$:

$$\int_Q u|_{Q_T} v_k\, dx = \int_{\Omega_T} f v_k\, dxdt + \int_Q \varphi v_k\, dx - \int_{\Omega_T} \nabla u \nabla v_k\, dxdt \quad (k = 1, 2 \ldots).$$

Since the system v_k is complete in $L_2(Q)$, it follows from Theorem 2.10 that the function $u|_{Q_T}$ as an element of the space $L_2(Q)$ is uniquely determined by these equalities. In addition, by virtue of Theorem 2.8 on the weak compactness of a closed ball in a Hilbert space, we have

$$\|u|_{Q_T}\|^2_{L_2(Q)} \leqslant d^2 \|f\|^2_{L_2(\Omega_T)} + \|\varphi\|^2_{L_2(Q)}.$$

In the special case $f = 0$, we get a particularly elegant estimate

$$\|u|_{Q_T}\|^2_{L_2(Q)} \leqslant \|\varphi\|^2_{L_2(Q)}.$$

It means that the linear operator transforming the initial function $\varphi \in L_2(Q)$ into a solution of the mixed problem for the homogeneous heat equation $u = u_\varphi$, considered for $T > 0$, is bounded in the space $L_2(Q)$ with norm not exceeding 1.

12 Cauchy problem for the wave equation

In this section, we follow the traditional path and consider classical solutions to the Cauchy problem. Of course, weak solutions can also be studied (see, e. g., [16], [11]).

A *classical solution* of the Cauchy problem for the wave equation is a function $u = u(x,t)$,

$$u \in C^2(\mathbb{R}^n \times \{t > 0\}) \cap C^1(\mathbb{R}^n \times \{t \geqslant 0\}),$$

satisfying the wave equation

$$\Box u := u_{tt} - \Delta u = f(x,t) \quad (x \in \mathbb{R}^n,\ t > 0) \tag{12.1}$$

in the half-space $\{t > 0\}$ of the space $\mathbb{R}^{n+1}_{x,t} = \{(x,t) : x \in \mathbb{R}^n, t \in \mathbb{R}\}$ and the initial conditions

$$u|_{t=0} = \varphi(x), \quad u_t|_{t=0} = \psi(x) \quad (x \in \mathbb{R}^n) \tag{12.2}$$

on the hyperplane $\{t = 0\}$.

Unique solvability of the Cauchy problem

Theorem 12.1. *Problem* (12.1), (12.2) *has at most one classical solution.*

Proof. Fixing a point $(\xi,\tau) \in \mathbb{R}^{n+1}$, $\tau > 0$, consider the cone

$$K_{\xi,\tau} = \{(x,t) \in \mathbb{R}^{n+1} : |x - \xi| < \tau - t,\ 0 < t < \tau\}.$$

Denote by

$$\Gamma_{\xi,\tau} = \{(x,t) \in \mathbb{R}^{n+1} : |x - \xi| = \tau - t,\ 0 < t < \tau\}$$

its lateral surface and by $R(h) = K_{\xi,\tau} \cap \{t = h\}$ the cross-section of $K_{\xi,\tau}$ by the hyperplane $\{t = h\}$, $0 < h < \tau$. We also need the conical surface

$$M(\varepsilon,h) = \Gamma_{\xi,\tau} \cap \{\varepsilon < t < h\} \quad (0 < \varepsilon < h < \tau)$$

and the truncated cone

$$V(\varepsilon,h) = K_{\xi,\tau} \cap \{\varepsilon < t < h\},$$

bounded by surfaces $\overline{R(\varepsilon)}$, $\overline{R(h)}$ and $M(\varepsilon,h)$.

We now need to show that if $u \in C^2(\mathbb{R}^n \times \{t > 0\}) \cap C^1(\mathbb{R}^n \times \{t \geqslant 0\})$ is a solution to the homogeneous problem

$$\Box u = 0 \quad (x \in \mathbb{R}^n,\ t > 0),$$
$$u(x,0) = u_t(x,0) = 0 \quad (x \in \mathbb{R}^n),$$

then this function is identically zero. For this purpose, we first multiply the homogeneous wave equation by $2u_t$, writing

$$2u_t u_{tt} - 2\sum_{i=1}^{n} u_t u_{x_i x_i} = 0,$$

and, after an obvious transformation, we get

$$(u_t^2 + |\nabla u|^2)_t - 2\sum_{i=1}^{n} (u_t u_{x_i})_{x_i} = 0$$

for $x \in \mathbb{R}^n$, $t > 0$. After this, the obtained equality is integrated over $V(\varepsilon, h)$:

$$\int\limits_{V(\varepsilon,h)} \left((u_t^2 + |\nabla u|^2)_t - 2\sum_{i=1}^{n} (u_t u_{x_i})_{x_i} \right) dxdt = 0. \tag{12.3}$$

If $\nu = \nu(x,t)$ is the unit outer normal vector to the smooth parts of the boundary of the domain $V(\varepsilon, h)$, then obviously

$$\nu|_{R(\varepsilon)} = (0,-1), \quad \nu|_{R(h)} = (0,1), \quad \nu|_{M(\varepsilon,h)} = \frac{1}{\sqrt{2}}\left(\frac{x-\xi}{|x-\xi|}, 1 \right).$$

Therefore, integrating (12.3) by parts, we have

$$\begin{aligned} \frac{1}{\sqrt{2}} \int\limits_{M(\varepsilon,h)} \left(u_t^2 + |\nabla u|^2 - 2u_t \nabla u \frac{x-\xi}{|x-\xi|} \right) dS_{x,t} \; + \\ + \int\limits_{R(h)} (u_t^2 + |\nabla u|^2)\, dx - \int\limits_{R(\varepsilon)} (u_t^2 + |\nabla u|^2)\, dx = 0. \end{aligned} \tag{12.4}$$

Due to the obvious inequality

$$u_t^2 + |\nabla u|^2 - 2u_t \nabla u \frac{x-\xi}{|x-\xi|} \geqslant u_t^2 + |\nabla u|^2 - 2|u_t||\nabla u| = (|u_t| - |\nabla u|)^2,$$

the first integral in (12.4) is nonnegative, and thus

$$\int\limits_{R(h)} (u_t^2 + |\nabla u|^2)\, dx \leqslant \int\limits_{R(\varepsilon)} (u_t^2 + |\nabla u|^2)\, dx.$$

It remains to pass to the limit as $\varepsilon \to 0$ to obtain

$$\int_{R(h)} (u_t^2 + |\nabla u|^2)\,dx = 0$$

for $0 < h < \tau$. Consequently, all first-order partial derivatives of the function u are equal to zero in $K_{\xi,\tau}$, and the function itself vanishes in $K_{\xi,\tau}$. □

Remark 12.1. The conical surface

$$\Gamma_{\xi,\tau} = \{(x,t) \in \mathbb{R}^{n+1} : |x-\xi| = \tau - t,\ 0 < t < \tau\}$$

with vertex (ξ,τ) is called the *characteristic surface* for the wave equation. It plays an important role in the study of the wave equation.

Solution of the Cauchy problem for dimension $n = 3$. Kirchhoff's formula

The solution formula to the Cauchy problem for the wave equation depends on the dimension n of the spatial variable x. At the beginning of this section, we consider a construction that is useful in all calculations related to the solution of the wave equation in the case $n = 3$. Given a function $g(x)$ of the variable $x \in \mathbb{R}^3$, we construct a function $u_g(x,t)$ of the two variables $x \in \mathbb{R}^3$ and $t > 0$ by the formula

$$u_g(x,t) = \frac{1}{4\pi t} \int_{|\xi-x|=t} g(\xi)\,dS_\xi. \tag{12.5}$$

The change $\xi = x + t\eta$ of variables in the integral brings the expression for $u_g(x,t)$ to the form

$$u_g(x,t) = \frac{t}{4\pi} \int_{|\eta|=1} g(x+t\eta)\,dS_\eta. \tag{12.6}$$

Lemma 12.1. *Let $g \in C^k(\mathbb{R}^3)$, $k \geqslant 2$. Then the function u_g belongs to the space $C^k(\mathbb{R}^3 \times \{t \geqslant 0\})$ and satisfies the relations*

$$u_g(x,0) = 0, \tag{12.7}$$

$$(u_g)_t(x,0) = g(x), \tag{12.8}$$

and

$$\Delta u_g(x,0) = 0 \tag{12.9}$$

for all $x \in \mathbb{R}^3$, as well as the homogeneous wave equation

$$\Box u_g = 0 \tag{12.10}$$

for all $x \in \mathbb{R}^3$, $t \geqslant 0$.

Proof. The fact that the function $u_g(x,t)$ is k-times continuously differentiable on the set $\{x \in \mathbb{R}^3, t \geqslant 0\}$ and equal to zero for $t = 0$ is an obvious consequence of representation (12.6). Moreover, we have

$$\Delta u_g(x,t) = \frac{t}{4\pi} \int\limits_{|\eta|=1} \Delta g(x + t\eta)\, dS_\eta \tag{12.11}$$

and

$$\begin{aligned}(u_g)_t(x,t) &= \frac{1}{4\pi} \int\limits_{|\eta|=1} g(x + t\eta)\, dS_\eta + \frac{t}{4\pi} \int\limits_{|\eta|=1} \eta \nabla g(x + t\eta)\, dS_\eta = \\ &= \frac{1}{4\pi} \int\limits_{|\eta|=1} g(x + t\eta)\, dS_\eta + \frac{t}{4\pi} \int\limits_{|\eta|=1} \frac{\partial g(x + t\eta)}{\partial \nu_\eta}\, dS_\eta,\end{aligned} \tag{12.12}$$

which immediately implies the relations $\Delta u_g(x,0) = 0$ and

$$(u_g)_t(x,0) = \frac{1}{4\pi} \int\limits_{|\eta|=1} g(x)\, dS_\eta = g(x).$$

Equalities (12.7)–(12.9) are now proven.

Returning to the former variable ξ in the integrals on the right-hand side of (12.11) and (12.12) and applying Green's formula, we obtain

$$\begin{aligned}(u_g)_t(x,t) &= \frac{1}{4\pi t^2} \int\limits_{|\xi - x|=t} g(\xi)\, dS_\xi + \frac{1}{4\pi t} \int\limits_{|\xi - x|=t} \frac{\partial g}{\partial \nu_\xi}\, dS_\xi = \\ &= \frac{1}{4\pi t^2} \int\limits_{|\xi - x|=t} g(\xi)\, dS_\xi + \frac{1}{4\pi t} \int\limits_{|\xi - x|<t} \Delta g(\xi)\, d\xi,\end{aligned} \tag{12.13}$$

while

$$\Delta u_g(x,t) = \frac{1}{4\pi t} \int\limits_{|\xi - x|=t} \Delta g(\xi)\, dS_\xi. \tag{12.14}$$

Denote

$$I_g(x,t) = \frac{1}{4\pi} \int\limits_{|\xi - x|<t} \Delta g(\xi)\, d\xi = \frac{1}{4\pi} \int\limits_0^t d\rho \int\limits_{|\xi - x|=\rho} \Delta g(\xi)\, dS_\xi.$$

Then

$$(I_g)_t(x,t) = \frac{1}{4\pi} \int\limits_{|\xi - x|=t} \Delta g(\xi)\, dS_\xi.$$

Equality (12.14) can now be represented in the form

$$\Delta u_g = \frac{(I_g)_t}{t}.$$

On the other hand, (12.13) gives

$$(u_g)_t = \frac{u_g + I_g}{t}.$$

Therefore,

$$(u_g)_{tt} = -\frac{u_g + I_g}{t^2} + \frac{(u_g)_t}{t} + \frac{(I_g)_t}{t} = -\frac{u_g + I_g}{t^2} + \frac{u_g + I_g}{t^2} + \frac{(I_g)_t}{t} = \frac{(I_g)_t}{t} = \Delta u_g.$$

Now, formula (12.10) has been proven. □

Lemma 12.1 means that the function $u_g(x,t)$ is a (classical) solution to the initial value problem

$$\Box u(x,t) = 0 \quad (x \in \mathbb{R}^3,\ t > 0),$$
$$u|_{t=0} = 0, \quad u_t|_{t=0} = g(x) \quad (x \in \mathbb{R}^3).$$

Note that for $k \geqslant 3$ the function $(u_g)_t(x,t)$ from $C^{k-1}(\mathbb{R}^3 \times \{t \geqslant 0\})$ is a solution to the initial value problem

$$\Box v(x,t) = 0 \quad (x \in \mathbb{R}^3,\ t > 0),$$
$$v|_{t=0} = g(x), \quad v_t|_{t=0} = 0 \quad (x \in \mathbb{R}^3).$$

Indeed, $\Box (u_g)_t = (\Box u_g)_t = 0$, and relation (12.8) provides the first of the initial conditions for $v = (u_g)_t$: $v(x,0) = (u_g)_t(x,0) = g(x)$. Moreover, from (12.10), we have

$$v_t(x,t) = (u_g)_{tt}(x,t) = \Delta u_g(x,t) \quad (x \in \mathbb{R}^3,\ t \geqslant 0),$$

whence it follows that $v_t(x,0) = 0$ by (12.9).

The reasoning above, together with the linearity of the problem, suggests that for $\varphi \in C^3(\mathbb{R}^3)$ and $\psi \in C^2(\mathbb{R}^3)$ the function $(u_\varphi)_t(x,t) + u_\psi(x,t)$ satisfies the initial value problem

$$\Box u(x,t) = 0 \quad (x \in \mathbb{R}^3,\ t > 0),$$
$$u|_{t=0} = \varphi(x), \quad u_t|_{t=0} = \psi(x) \quad (x \in \mathbb{R}^3)$$

in the case of the homogeneous wave equation.

Passing to the nonhomogeneous wave equation, we seek a particular solution to the equation $\Box u(x,t) = f(x,t)$ in the integral form

$$\int_0^t u_f(x, t-\tau, \tau)\, d\tau,$$

where

$$u_f(x,t,\tau) = \frac{1}{4\pi t} \int\limits_{|\xi - x| = t} f(\xi, \tau)\, dS_\xi.$$

The following "τ-dependent" generalization of Lemma 12.1 is obvious: if the function $f(x,\tau)$ is continuous with respect to both variables x, τ and k-times continuously differentiable with respect to x on the set $\{x \in \mathbb{R}^3, \tau \geqslant 0\}$, then the function $u_f(x,t,\tau)$ is continuous with respect to x, t, τ and k-times continuously differentiable with respect to x, t on the set $\{x \in \mathbb{R}^3, t \geqslant 0, \tau \geqslant 0\}$. This function satisfies the equalities $u_f(x, 0, \tau) = 0$, $(u_f)_t(x, 0, \tau) = f(x, \tau)$ and $\Delta u_f(x, 0, \tau) = 0$ for all $x \in \mathbb{R}^3$, $\tau \geqslant 0$, and the homogeneous wave equation

$$(u_f)_{tt}(x,t,\tau) = \Delta u_f(x,t,\tau)$$

in $\{x \in \mathbb{R}^3, t \geqslant 0, \tau \geqslant 0\}$. As before, it is assumed that $k \geqslant 2$.

Under the conditions formulated above, set

$$w(x,t) = \int_0^t u_f(x, t-\tau, \tau)\, d\tau. \tag{12.15}$$

It is easy to see that the functions $w(x,t)$ and

$$w_t(x,t) = u_f(x,0,t) + \int_0^t (u_f)_t(x, t-\tau, \tau)\, d\tau = \int_0^t (u_f)_t(x, t-\tau, \tau)\, d\tau$$

vanish at $t = 0$. Moreover,

$$\begin{aligned} w_{tt}(x,t) = (u_f)_t(x,0,t) + \int_0^t (u_f)_{tt}(x, t-\tau, \tau)\, d\tau = f(x,t) + {} \\ + \int_0^t \Delta u_f(x, t-\tau, \tau)\, d\tau = f(x,t) + \Delta w(x,t) \end{aligned}$$

for all $x \in \mathbb{R}^3$, $t \geqslant 0$. This shows that function (12.15) satisfies the initial value problem

$$\begin{aligned} &\Box\, w(x,t) = f(x,t) \quad (x \in \mathbb{R}^3,\ t > 0), \\ &w|_{t=0} = 0, \quad w_t|_{t=0} = 0 \quad (x \in \mathbb{R}^3). \end{aligned}$$

Putting it all together, we see that the function

$$u(x,t) = (u_\varphi)_t(x,t) + u_\psi(x,t) + \int_0^t u_f(x, t-\tau, \tau)\, d\tau$$

is a solution to the initial value problem

$$\Box u(x,t) = f(x,t) \quad (x \in \mathbb{R}^3,\ t > 0), \tag{12.16}$$

$$u|_{t=0} = \varphi(x), \quad u_t|_{t=0} = \psi(x) \quad (x \in \mathbb{R}^3). \tag{12.17}$$

Setting $\rho = t - \tau$, one can transform the integral in (12.15) as follows:

$$\int_0^t u_f(x, t-\tau, \tau)\, d\tau = \int_0^t \frac{d\tau}{4\pi(t-\tau)} \int_{|\xi - x| = t-\tau} f(\xi, \tau)\, dS_\xi =$$

$$= \int_0^t \frac{d\rho}{4\pi\rho} \int_{|\xi-x|=\rho} f(\xi, t-\rho)\, dS_\xi = \frac{1}{4\pi}\int_0^t d\rho \int_{|\xi-x|=\rho} \frac{f(\xi, t-|\xi-x|)}{|\xi-x|}\, dS_\xi =$$

$$= \frac{1}{4\pi}\int_{|\xi-x|<t} \frac{f(\xi, t-|\xi-x|)}{|\xi-x|}\, d\xi.$$

Let us formalize the reasoning above.

Theorem 12.2. *Let the functions $\varphi(x)$ and $\psi(x)$ be such that $\varphi \in C^3(\mathbb{R}^3)$, $\psi \in C^2(\mathbb{R}^3)$, and the function $f(x,t)$ be continuous and twice continuously differentiable with respect to the spatial variables x on the set $\mathbb{R}^3 \times \{t \geqslant 0\}$. Then there exists a classical solution $u(x,t)$ to problem* (12.16) *and* (12.17). *This solution belongs to the space $C^2(\mathbb{R}^3 \times \{t \geqslant 0\})$ and is given by the Kirchhoff formula*

$$u(x,t) = \frac{\partial}{\partial t}\left(\frac{1}{4\pi t}\int_{|\xi-x|=t} \varphi(\xi)\, dS_\xi\right) + \frac{1}{4\pi t}\int_{|\xi-x|=t} \psi(\xi)\, dS_\xi + \\ + \frac{1}{4\pi}\int_{|\xi-x|<t} \frac{f(\xi, t-|\xi-x|)}{|\xi-x|}\, d\xi. \tag{12.18}$$

Remark 12.2. Let us make one important observation related to formula (12.18). Since

$$\frac{\partial}{\partial t}\left(\frac{1}{4\pi t}\int_{|\xi-x|=t} \varphi(\xi)\, dS_\xi\right) = \frac{1}{4\pi t^2}\int_{|\xi-x|=t} \varphi(\xi)\, dS_\xi + \frac{1}{4\pi t}\int_{|\xi-x|=t} \frac{\partial\varphi}{\partial\nu_\xi}\, dS_\xi,$$

it becomes clear from (12.18) that the solution $u(x,t)$ of the Cauchy problem depends at each point (x,t) on the values of the function f on the characteristic (conical) surface

$\Gamma_{x,t}$, and on the values of the initial value functions (together with the normal derivative of the first initial function) on the sphere

$$S_{x,t} = \{\xi \in \mathbb{R}^3 : |\xi - x| = t\}$$

at the base of this cone. In particular, we have $u(x,t) = 0$ if $f = 0$ on $\Gamma_{x,t}$ and $\varphi = \partial\varphi/\partial\nu = \psi = 0$ on $S_{x,t}$. The sphere $S_{x,t}$ is therefore called the *region of influence* of the initial conditions on the solution of the Cauchy problem at the point (x,t). We can say that the solution "knows" at the point x what happened the time t before the current moment only at points located at the distance t from the point x. In other words, according to the model under consideration, disturbances propagate in the medium at unit speed.

Moreover, assuming the medium to be initially in a state of equilibrium outside the ball $\{|x-x^0| < R\}$, we immediately obtain that over time t (and in the absence of external forces) the disturbance takes place only at points of the spherical layer

$$t - R < |x - x^0| < t + R.$$

This fact can be expressed differently: the disturbance is present at a point $x \in \mathbb{R}^3$ only in the time interval $|x - x^0| - R < t < |x - x^0| + R$.

Solution of the Cauchy problem for dimensions $n = 1$ and $n = 2$. D'Alembert and Poisson formulas

Knowing how the solution to the Cauchy problem looks like for $n = 3$, we can obtain the corresponding representations in lower dimensions using the so-called *descent method*. Consider again the three-dimensional Cauchy problem, this time assuming that the functions $\varphi(x_1, x_2, x_3)$, $\psi(x_1, x_2, x_3)$ and $f(x_1, x_2, x_3, t)$ do not depend on the coordinate x_3. Under the previous assumptions regarding the smoothness of these functions, we represent the solution of this problem by formula (12.18).

Remark 12.3. In the case where the function depends only on two of the three coordinates, $g = g(x_1, x_2)$, it is convenient to transform the corresponding integral over the sphere to the double integral,

$$\int\limits_{|\xi-x|=t} g(\xi)\, dS_\xi = 2 \int\limits_{|\xi-x|=t,\, \xi_3>x_3} g(\xi)\, dS_\xi = \\ = 2t \int\limits_{(\xi_1-x_1)^2+(\xi_2-x_2)^2<t^2} \frac{g(\xi_1, \xi_2)\, d\xi_1 d\xi_2}{\sqrt{t^2 - (\xi_1 - x_1)^2 - (\xi_2 - x_2)^2}},$$

since

$$dS_\xi = \frac{|\xi - x|\, d\xi_1 d\xi_2}{\xi_3 - x_3} = \frac{t\, d\xi_1 d\xi_2}{\sqrt{t^2 - (\xi_1 - x_1)^2 - (\xi_2 - x_2)^2}}$$

on the upper half of the sphere.

Taking this into account, we obtain the following representation for the function on the right-hand side of formula (12.18):

$$\begin{aligned}
&\frac{\partial}{\partial t}\left(\frac{1}{2\pi} \int\limits_{(\xi_1-x_1)^2+(\xi_2-x_2)^2<t^2} \frac{\varphi(\xi_1, \xi_2)\, d\xi_1 d\xi_2}{\sqrt{t^2 - (\xi_1 - x_1)^2 - (\xi_2 - x_2)^2}}\right) + \\
&+ \frac{1}{2\pi} \int\limits_{(\xi_1-x_1)^2+(\xi_2-x_2)^2<t^2} \frac{\psi(\xi_1, \xi_2)\, d\xi_1 d\xi_2}{\sqrt{t^2 - (\xi_1 - x_1)^2 - (\xi_2 - x_2)^2}} + \\
&+ \frac{1}{2\pi} \int_0^t d\rho \int\limits_{(\xi_1-x_1)^2+(\xi_2-x_2)^2<\rho^2} \frac{f(\xi_1, \xi_2, t-\rho)\, d\xi_1 d\xi_2}{\sqrt{\rho^2 - (\xi_1 - x_1)^2 - (\xi_2 - x_2)^2}}.
\end{aligned}$$

An important observation is that this function is also independent of the x_3 coordinate, and thus satisfies the two-dimensional Cauchy problem

$$\Box u = f(x_1, x_2, t) \quad ((x_1, x_2) \in \mathbb{R}^2, t > 0), \tag{12.19}$$

$$u|_{t=0} = \varphi(x_1, x_2), \quad u_t|_{t=0} = \psi(x_1, x_2) \quad ((x_1, x_2) \in \mathbb{R}^2). \tag{12.20}$$

The theorem below repeats Theorem 12.2 in dimension $n = 2$. Here, we return to the shortened notation $x = (x_1, x_2)$ and $\xi = (\xi_1, \xi_2)$.

Theorem 12.3. *Let the functions $\varphi(x)$ and $\psi(x)$ belong to the spaces $C^3(\mathbb{R}^2)$ and $C^2(\mathbb{R}^2)$, respectively, and the function $f(x, t)$ be continuous and twice continuously differentiable with respect to the variables x on the set $\mathbb{R}^2 \times \{t \geqslant 0\}$. Then there exists a classical solution $u(x, t)$ to problem* (12.19) *and* (12.20)*. This solution belongs to $C^2(\mathbb{R}^2 \times \{t \geqslant 0\})$ and is given by the Poisson formula*

$$\begin{aligned}
u(x, t) = &\frac{\partial}{\partial t}\left(\frac{1}{2\pi} \int\limits_{|\xi-x|<t} \frac{\varphi(\xi)\, d\xi}{\sqrt{t^2 - |\xi - x|^2}}\right) + \frac{1}{2\pi} \int\limits_{|\xi-x|<t} \frac{\psi(\xi)\, d\xi}{\sqrt{t^2 - |\xi - x|^2}} + \\
&+ \frac{1}{2\pi} \int_0^t d\rho \int\limits_{|\xi-x|<\rho} \frac{f(\xi, t-\rho)\, d\xi}{\sqrt{\rho^2 - |\xi - x|^2}}.
\end{aligned} \tag{12.21}$$

Remark 12.4. There is a fundamental difference between the Kirchhoff and the Poisson formulas. In the latter case, the function f is integrated over the entire cone $K_{x,t}$ and not just over the conical surface $\Gamma_{x,t}$. The initial value functions are integrated over the whole circle

$$D_{x,t} = \{\xi \in \mathbb{R}^2 : |\xi - x| < t\}$$

and not along its boundary $S_{x,t}$. Therefore, the conditions

$$f = 0 \quad \text{on } \Gamma_{x,t}, \quad \varphi = \frac{\partial \varphi}{\partial \nu} = \psi = 0 \quad \text{on } S_{x,t}$$

are no longer sufficient to guarantee $u(x,t) = 0$. Instead, we should require that the function f vanish everywhere in $K_{x,t}$ and the initial value functions vanish on $D_{x,t}$. The circle $D_{x,t}$ is the *region of influence* of the initial data on the solution at the point (x,t). A disturbance localized initially in the circle $|x - x^0| < R$ reaches a point y at the time $t = |y - x^0| - R$, but unlike the case of $\mathbb{R}^3$, remains at this point, generally speaking, for an arbitrarily long time.

The corresponding representation of the solution to the Cauchy problem for the wave equation in the case $n = 1$ looks much simpler. The solution $u(x,t)$ to the Cauchy problem

$$u_{tt} - u_{xx} = f(x,t) \quad (x \in \mathbb{R},\ t > 0),$$
$$u|_{t=0} = \varphi(x), \quad u_t|_{t=0} = \psi(x) \quad (x \in \mathbb{R})$$

is given by the *D'Alembert formula*

$$u(x,t) = \frac{\varphi(x+t) + \varphi(x-t)}{2} + \frac{1}{2}\int_{x-t}^{x+t} \psi(\xi)\,d\xi + \frac{1}{2}\int_0^t d\tau \int_{x-(t-\tau)}^{x+(t-\tau)} f(\xi,\tau)\,d\xi. \tag{12.22}$$

We leave the derivation of this formula and its analysis to the reader, noting only that it gives a solution to the Cauchy problem with less constraints on the function on the right-hand side of the equation and the initial data.

13 Cauchy problem for the heat equation

Unique solvability of the Cauchy problem

We start this section with the derivation of the *weak maximum principle* for the heat equation. In particular, a uniqueness of the classical solution bounded as $|x| \to \infty$ follows from this principle.

Given numbers $R > 0$, $\tau > 0$, consider the cylinder

$$\Omega_\tau = \{(x,t) \in \mathbb{R}^{n+1} : |x| < R,\ 0 < t < \tau\}.$$

We denote by $D_0 = \{(x,0) : |x| < R\}$ and $D_\tau = \{(x,\tau) : |x| < R\}$ its lower and upper bases, and by

$$S_\tau = \{(x,t) : |x| = R,\ 0 < t < \tau\}$$

its lateral surface.

Lemma 13.1. *Let the real-valued function $u(x,t)$ defined in $\overline{\Omega}_\tau$ belong to the space $C^{2,1}(\Omega_\tau \cup D_\tau) \cap C(\overline{\Omega}_\tau)$ and satisfy the homogeneous heat equation in $\Omega_\tau \cup D_\tau$, $u_t = \Delta u$. Then the smallest and largest values of u are achieved at $\overline{S}_\tau \cup D_0$, i. e.,*

$$\min_{\overline{S}_\tau \cup D_0} u \leqslant u(x,t) \leqslant \max_{\overline{S}_\tau \cup D_0} u \tag{13.1}$$

for all $(x,t) \in \overline{\Omega}_\tau$.

Proof. Let us prove the inequality

$$u(x,t) \geqslant \min_{\overline{S}_\tau \cup D_0} u \quad ((x,t) \in \overline{\Omega}_\tau),$$

the other part of (13.1) can be proved in a similar way.

By subtracting a sufficiently large constant from u if necessary, we assume without loss of generality that the function u is strictly negative in $\overline{\Omega}$. Set $u = e^{\gamma t} v$ with an arbitrary positive parameter γ, then

$$v_t - \Delta v = -\gamma e^{-\gamma t} u > 0 \tag{13.2}$$

in $\Omega_\tau \cup D_\tau$. Therefore, v cannot reach its minimum at $\Omega_\tau \cup D_\tau$. Indeed, if $(x^0, t^0) \in \Omega_\tau \cup D_\tau$ is a point of minimum for v, then

$$v_t(x^0, t^0) \leqslant 0, \quad v_{x_i x_i}(x^0, t^0) \geqslant 0 \quad (i = 1, \dots, n),$$

which contradicts (13.2). Consequently, the smallest value of the function v on the set $\overline{\Omega}_\tau$ is achieved at $\overline{S}_\tau \cup D_0$, i. e.,

$$e^{-\gamma t} u(x,t) \geqslant \min_{\overline{S}_\tau \cup D_0} e^{-\gamma t} u \quad ((x,t) \in \overline{\Omega}_\tau) \tag{13.3}$$

for all $\gamma > 0$. Denote

$$m(\gamma) = \min_{\overline{S}_\tau \cup D_0} e^{-\gamma t} u, \quad \gamma \geqslant 0.$$

Then we have $e^{-\gamma t} u(x,t) \geqslant m(\gamma)$ for $\gamma \geqslant 0$ on $\overline{S}_\tau \cup D_0$. In particular, $u(x,t) \geqslant m(0)$ and $e^{-\gamma t} u(x,t) \geqslant e^{-\gamma t} m(0) \geqslant e^{-\gamma \tau} m(0)$ on $\overline{S}_\tau \cup D_0$, which means $m(\gamma) \geqslant e^{-\gamma \tau} m(0)$. From (13.3), it now follows that

$$u(x,t) \geqslant e^{\gamma(t-\tau)} m(0) \quad (\gamma > 0,\ (x,t) \in \overline{\Omega}_\tau).$$

Passing to the limit as $\gamma \to 0$ in this relation, we arrive at the required inequality. □

Theorem 13.1. *Let a real-valued function $u(x,t)$ defined in the layer $\mathbb{R}^n \times [0,T)$ belong to the space $C^{2,1}(\mathbb{R}^n \times (0,T)) \cap C(\mathbb{R}^n \times [0,T))$ and satisfy the homogeneous heat equation $u_t = \Delta u$ in $\mathbb{R}^n \times (0,T)$. If, in addition, $u(x,t)$ is bounded in the layer $\mathbb{R}^n \times (0,T)$, then at each point $(x,t) \in \mathbb{R}^n \times [0,T)$ the following inequality holds:*

$$\inf_{\mathbb{R}^n} u(x,0) \leqslant u(x,t) \leqslant \sup_{\mathbb{R}^n} u(x,0). \tag{13.4}$$

Proof. Choosing an arbitrary $\tau \in (0,T)$, we prove inequality (13.4) for all points of the layer $\mathbb{R}^n \times [0,\tau]$. We restrict ourselves again to proving the left-hand side of the inequality. Let us denote $m_0 = \inf_{\mathbb{R}^n} u(x,0)$ and consider the function

$$v_\varepsilon(x,t) = u(x,t) - m_0 + \varepsilon(|x|^2 + 2nt)$$

depending on the positive parameter ε. This function also satisfies the homogeneous heat equation in $\mathbb{R}^n \times (0,T)$ and is such that $v_\varepsilon(x,0) \geqslant 0$.

In addition, for a fixed ε, we choose a large enough $R > 0$ such that $v_\varepsilon(x,t) \geqslant 0$ for $|x| \geqslant R$, $t \in [0,\tau]$. This can be done due to the boundedness of u in $\mathbb{R}^n \times [0,\tau]$. Applying Lemma 13.1 to the function v_ε in $\Omega_\tau = \{(x,t) : |x| < R, 0 < t < \tau\}$, we see that this function is nonnegative in $\overline{\Omega}_\tau$ and, therefore, in the entire layer $\mathbb{R}^n \times [0,\tau]$. Due to the arbitrariness of ε, we obtain in the limit as $\varepsilon \to 0$ that $u(x,t) \geqslant m_0$ in $\mathbb{R}^n \times [0,\tau]$. The proof is complete. The same approach can be applied to prove the right-hand side of (13.4). □

By a *classical solution* of the Cauchy problem in $\mathbb{R}^n \times (0,T)$, we understand a function $u(x,t)$ from the class

$$C^{2,1}(\mathbb{R}^n \times (0,T)) \cap C(\mathbb{R}^n \times [0,T)),$$

satisfying the heat equation

$$u_t - \Delta u = f(x,t) \quad (x \in \mathbb{R}^n, 0 < t < T) \tag{13.5}$$

and taking specified values

$$u|_{t=0} = \varphi(x) \quad (x \in \mathbb{R}^n) \tag{13.6}$$

on the hyperplane $t = 0$. Here, $f \in C(\mathbb{R}^n \times (0,T))$ and $\varphi \in C(\mathbb{R}^n)$.

Theorem 13.2. *Problem* (13.5), (13.6) *cannot have more than one classical solution bounded in the layer* $\mathbb{R}^n \times (0,T)$.

Proof. If u_1 and u_2 are classical solutions to problem (13.5) and (13.6) bounded in the layer $\mathbb{R}^n \times (0,T)$, then their difference $u = u_1 - u_2$ satisfies all the conditions of Theorem 13.1. Moreover, u vanishes at $t = 0$. By Theorem 13.1, u is identically zero in $\mathbb{R}^n \times [0,T)$. □

Remark 13.1. The boundedness condition is essential in Theorem 13.2 since the homogeneous Cauchy problem has nontrivial classical solutions growing as $|x| \to \infty$. The corresponding example was constructed by A. N. Tikhonov.[1]

Poisson integral for the heat equation

We begin this section with a formal derivation of the solution to the Cauchy problem. Next, the conditions on the functions $f(x,t)$ and $\varphi(x)$ are indicated under which the resulting formula gives a classical solution to the problem.

For simplicity, set $f = 0$, i. e., let us first solve the homogeneous heat equation. To do this, we use the Fourier transform in spatial variables x,

$$u(x,t) \longmapsto \hat{u}(\xi,t) = (2\pi)^{-n/2} \int_{\mathbb{R}^n} e^{-i(\xi,x)} u(x,t)\,dx, \quad \xi \in \mathbb{R}^n.$$

Since $\widehat{(u_t)} = \hat{u}_t$ and $\widehat{(\Delta u)} = -|\xi|^2 \hat{u}$, we obtain the ordinary differential equation

$$\hat{u}_t(\xi,t) + |\xi|^2 \hat{u}(\xi,t) = 0 \quad (t > 0) \tag{13.7}$$

with the parameter $\xi \in \mathbb{R}^n$ for the Fourier image $\hat{u}(\xi,t)$ of the desired function $u(x,t)$. The initial condition $u|_{t=0} = \varphi(x)$ is transformed into

$$\hat{u}(\xi,0) = \hat{\varphi}(\xi). \tag{13.8}$$

From (13.7) and (13.8), it immediately follows that

$$\hat{u}(\xi,t) = \hat{\varphi}(\xi) e^{-|\xi|^2 t}.$$

Recall another fundamental property of the Fourier transform,

$$(2\pi)^{-n/2} \widehat{(v * w)} = \hat{v}\hat{w},$$

where $*$ denotes convolution,

$$(v * w)(x) = \int_{\mathbb{R}^n} v(x-y) w(y)\,dy.$$

Then we come to the conclusion that the solution to the original Cauchy problem is the convolution of the initial value function with the Fourier preimage

1 A. Tikhonov, "Théorèmes d'unicité pour l'équation de la chaleur," Matem. Sat., 42:2 (1935), 199–216.

$$U(x,t) = (2\pi)^{-n} \int_{\mathbb{R}^n} e^{i(\xi,x)} e^{-|\xi|^2 t}\, d\xi \qquad (13.9)$$

of the function $e^{-|\xi|^2 t}$ multiplied by $(2\pi)^{-n/2}$. Let us compute the last integral, representable as the product

$$(2\pi)^{-n} \prod_{j=1}^{n} \int_{-\infty}^{+\infty} e^{i\mu x_j - \mu^2 t}\, d\mu$$

of one-dimensional integrals of the same type. The change $\mu = \eta/\sqrt{t}$ of the integration variable leads to

$$\int_{-\infty}^{+\infty} e^{i\mu x_j - \mu^2 t} d\mu = \frac{1}{\sqrt{t}} \int_{-\infty}^{+\infty} e^{-(\eta^2 - i x_j \eta/\sqrt{t})} d\eta = \frac{1}{\sqrt{t}} e^{-\frac{x_j^2}{4t}} \int_{-\infty}^{+\infty} e^{-(\eta - i\frac{x_j}{2\sqrt{t}})^2} d\eta.$$

Since the function e^{-z^2} of the complex variable z is analytic in a finite part of the complex plane, the following contour integral is equal to zero:

$$\int_{-R}^{R} e^{-\eta^2} d\eta + i \int_{0}^{-\frac{x_j}{2\sqrt{t}}} e^{-(i\sigma+R)^2} d\sigma - \int_{-R}^{R} e^{-(\eta - i\frac{x_j}{2\sqrt{t}})^2} d\eta - i \int_{0}^{-\frac{x_j}{2\sqrt{t}}} e^{-(i\sigma-R)^2} d\sigma = 0.$$

Passing to the limit as $R \to +\infty$ in this equality, we see that

$$\int_{-\infty}^{+\infty} e^{-(\eta - i\frac{x_j}{2\sqrt{t}})^2} d\eta = \int_{-\infty}^{+\infty} e^{-\eta^2} d\eta = \sqrt{\pi}$$

and, therefore, integral (13.9) is equal to

$$U(x,t) = \frac{1}{(2\sqrt{\pi t})^n} e^{-\frac{|x|^2}{4t}}.$$

The function $U(x,t)$ is called the *fundamental solution* of the heat equation. One could easily verify that it satisfies the homogeneous heat equation for $t > 0$.

Thus, we have obtained the (for now formal) representation

$$u(x,t) = \int_{\mathbb{R}^n} U(x-y,t)\varphi(y)\, dy = \frac{1}{(2\sqrt{\pi t})^n} \int_{\mathbb{R}^n} e^{-\frac{|x-y|^2}{4t}} \varphi(y)\, dy$$
$$(x \in \mathbb{R}^n,\ t > 0)$$

of the solution $u(x,t)$ to the Cauchy problem

$$u_t - \Delta u = 0 \quad (x \in \mathbb{R}^n, t > 0), \quad u(x,0) = \varphi(x) \quad (x \in \mathbb{R}^n).$$

The integral on the right-hand side of the obtained representation is called the *Poisson integral.*

Lemma 13.2. *Let a function $\varphi = \varphi(x)$ be continuous and bounded in $\mathbb{R}^n$. Then the function*

$$u_\varphi(x,t) = \begin{cases} \int_{\mathbb{R}^n} U(x-y,t)\varphi(y)\,dy, & t>0, \\ \varphi(x), & t=0 \end{cases}$$

is infinitely differentiable and satisfies the homogeneous heat equation $u_{\varphi t} = \Delta u_\varphi$ in the half-space $\mathbb{R}^n \times (0,+\infty)$. Moreover, it is continuous in $\mathbb{R}^n \times [0,+\infty)$ and $|u_\varphi(x,t)| \leqslant \sup|\varphi|$.

Proof. The smoothness of u_φ for $t > 0$ follows from the infinite differentiability of $\int_{\mathbb{R}^n} U(x-y,t)\varphi(y)\,dy$ under the integral sign since all the integrals $\int_{\mathbb{R}^n} \varphi(y) D^{\alpha,l}_{x,t} U(x-y,t)\,dy$ converge uniformly on compact subsets of $\mathbb{R}^n \times (0,+\infty)$. Moreover,

$$u_{\varphi t}(x,t) - \Delta u_\varphi(x,t) = \int\limits_{\mathbb{R}^n} \varphi(y)(U_t(x-y,t) - \Delta U(x-y,t))\,dy = 0$$

for $t > 0$.

In order to verify the continuity of u_φ up to the boundary of the half-space, we make the change of variables $y = x + 2\sqrt{t}\eta$ in the integral $\int_{\mathbb{R}^n} U(x-y,t)\varphi(y)\,dy$, bringing it to the form

$$u_\varphi(x,t) = \pi^{-n/2} \int\limits_{\mathbb{R}^n} e^{-|\eta|^2} \varphi(x + 2\sqrt{t}\eta)\,d\eta, \quad t>0.$$

For an arbitrary point $x^0 \in \mathbb{R}^n$, we write

$$u_\varphi(x,t) - \varphi(x^0) = \pi^{-n/2} \int\limits_{\mathbb{R}^n} e^{-|\eta|^2} [\varphi(x + 2\sqrt{t}\eta) - \varphi(x^0)]\,d\eta$$

and

$$\begin{aligned} |u_\varphi(x,t) - \varphi(x^0)| \leqslant 2\pi^{-n/2} \sup|\varphi| \int\limits_{|\eta|>N} e^{-|\eta|^2}\,d\eta \,+ \\ + \pi^{-n/2} \int\limits_{|\eta|<N} e^{-|\eta|^2} |\varphi(x + 2\sqrt{t}\eta) - \varphi(x^0)|\,d\eta. \end{aligned}$$

By taking the number N large enough, we ensure that the first integral on the right-hand side of this inequality is small. Due to the continuity of the function φ at the point x^0, the second integral is small if

$$|x - x^0| < \delta \quad \text{and} \quad t < \left(\frac{\delta}{2N}\right)^2$$

for sufficiently small δ. This means that the function $u_\varphi(x,t)$ is continuous for $t \geqslant 0$. The inequality $|u_\varphi(x,t)| \leqslant \sup|\varphi|$ is obvious for all $x \in \mathbb{R}^n$ and $t \geqslant 0$. □

Therefore, if the initial function $\varphi(x)$ is continuous and bounded, then the function $u_\varphi(x,t)$ is a classical solution to the Cauchy problem for the homogeneous heat equation.

Next, given a function $f(x,t)$ continuous and bounded in the layer $\mathbb{R}^n \times (0,T)$, consider the integral

$$u_f(x,t-s,s) = \int\limits_{\mathbb{R}^n} U(x-y,t-s)f(y,s)dy = \\ = \pi^{-n/2}\int\limits_{\mathbb{R}^n} e^{-|\eta|^2} f(x+2\sqrt{t-s}\,\eta,s)\,d\eta.$$

This integral has continuous derivatives of any order with respect to x and t and satisfies the homogeneous heat equation $u_{ft}(x,t-s,s) = \Delta u_f(x,t-s,s)$ for $x \in \mathbb{R}^n$, $t \in (0,T)$ and $s \in (0,t)$. In addition, the function $u_f(x,t-s,s)$ is continuous and bounded on the set $\{x \in \mathbb{R}^n, t \in (0,T), s \in (0,t]\}$, $|u_f(x,t-s,s)| \leqslant \sup|f(x,t)|$, and the following equality holds for $s = t$:

$$u_f(x,0,t) = f(x,t) \quad (x \in \mathbb{R}^n, t > 0)$$

(see proof of Lemma 13.2).

Now let us additionally assume that f has continuous and bounded derivatives $f_{x_i}(x,t)$ $(i = 1,\dots,n)$ in the layer $\mathbb{R}^n \times (0,T)$. Then the second-order derivatives of u_f with respect to the spatial variables and the first-order derivative with respect to t can be expressed as follows:

$$u_{fx_i}(x,t-s,s) = \pi^{-n/2}\int\limits_{\mathbb{R}^n} e^{-|\eta|^2} f_{x_i}(x+2\sqrt{t-s}\,\eta,s)\,d\eta = \\ = \frac{1}{2\pi^{n/2}\sqrt{t-s}}\int\limits_{\mathbb{R}^n} e^{-|\eta|^2}\big(f(x+2\sqrt{t-s}\,\eta,s)\big)_{\eta_i}\,d\eta = \\ = \frac{1}{\pi^{n/2}\sqrt{t-s}}\int\limits_{\mathbb{R}^n} \eta_i e^{-|\eta|^2} f(x+2\sqrt{t-s}\,\eta,s)\,d\eta$$

(the last equality is obtained with integration by parts),

$$u_{fx_ix_j}(x,t-s,s) = \frac{1}{\pi^{n/2}\sqrt{t-s}}\int\limits_{\mathbb{R}^n} \eta_i e^{-|\eta|^2} f_{x_j}(x+2\sqrt{t-s}\,\eta,s)\,d\eta,$$

$$u_{ft}(x,t-s,s) = \frac{1}{\pi^{n/2}\sqrt{t-s}}\int\limits_{\mathbb{R}^n} e^{-|\eta|^2}\sum_{i=1}^{n}\eta_i f_{x_i}(x+2\sqrt{t-s}\,\eta,s)\,d\eta = \\ = \Delta u_f(x,t-s,s) \quad (x \in \mathbb{R}^n, t \in (0,T), s \in (0,t)),$$

whence the estimates

$$|u_{ft}|,\ |u_{fx_i}|,\ |u_{fx_ix_j}| \leqslant C/\sqrt{t-s} \tag{13.10}$$

follow with a constant $C > 0$.

Lemma 13.3. *Let a function $f(x,t)$ and its partial derivatives f_{x_i} ($i = 1,\dots,n$) be continuous and bounded in the layer $\mathbb{R}^n \times (0,T)$. Then the integral*

$$I(x,t) = \int_0^t u_f(x, t-s, s)\, ds \quad (x \in \mathbb{R}^n, t \in [0,T))$$

is a classical solution to the Cauchy problem

$$I_t(x,t) - \Delta I(x,t) = f(x,t) \quad (x \in \mathbb{R}^n, t > 0), \qquad I(x,0) = 0,$$

and $|I(x,t)| \leqslant T \sup |f|$.

Proof. Continuity of $I(x,t)$ on the set $\mathbb{R}^n \times [0,T)$, the equality $I(x,0) = 0$, and the estimate $|I(x,t)| \leqslant T \sup |f|$ are obvious. Moreover, it follows from inequalities (13.10) and the properties of parameter-dependent improper integrals that the function I is twice continuously differentiable with respect to the variables $x_1,\dots,x_n$ on the set $\mathbb{R}^n \times (0,T)$ and

$$\Delta I(x,t) = \int_0^t \Delta u_f(x, t-s, s)\, ds.$$

It remains to verify that the integral $I(x,t)$ has a derivative with respect to t equal to $f(x,t) + \int_0^t \Delta u_f(x, t-s, s)\, ds$ for $t > 0$. Fix $(x,t) \in \mathbb{R}^n \times (0,T)$. For positive increments h of the argument t, we can write

$$\frac{I(x,t+h) - I(x,t)}{h} = \frac{1}{h} \int_t^{t+h} u_f(x, t+h-s, s)\, ds + \\ + \int_0^t \frac{u_f(x, t+h-s, s) - u_f(x, t-s, s)}{h}\, ds. \tag{13.11}$$

It follows now from the continuity of the function $u_f(x, t+h-s, s)$ with respect to $s \in [t, t+h]$ and the integral mean value theorem that the first term on the right-hand side of (13.11) tends to $u_f(x,0,t) = f(x,t)$ as $h \to +0$. In order to estimate the expression under the sign of the second integral, we use the Newton–Leibniz formula and inequalities (13.10):

$$\left|\frac{u_f(x,t+h-s,s)-u_f(x,t-s,s)}{h}\right| \leqslant \frac{1}{h}\int_t^{t+h} |u_{ft}(x,\tau-s,s)|\,d\tau \leqslant$$

$$\leqslant \frac{C}{h}\int_t^{t+h} \frac{d\tau}{\sqrt{\tau-s}} \leqslant \frac{C_1}{\sqrt{t-s}}$$

($C_1 > 0$ does not depend on h). Therefore, Lebesgue's dominated convergence theorem is applicable to the second integral on the right-hand side of (13.11), according to which this integral tends to

$$\int_0^t u_{ft}(x,t-s,s)\,ds = \int_0^t \Delta u_f(x,t-s,s)\,ds = \Delta I(x,t)$$

as $h \to +0$. Therefore,

$$\lim_{h\to+0} \frac{I(x,t+h)-I(x,t)}{h} = f(x,t) + \Delta I(x,t).$$

Similar reasoning shows that

$$\lim_{h\to-0} \frac{I(x,t+h)-I(x,t)}{h} = -\lim_{h\to-0} \frac{1}{h}\int_{t+h}^{t} u_f(x,t-s,s)\,ds +$$

$$+\lim_{h\to-0} \int_0^{t+h} \frac{u_f(x,t+h-s,s)-u_f(x,t-s,s)}{h}\,ds = f(x,t)+\Delta I(x,t).$$

The lemma is now proven. □

Combining Lemmas 13.2 and 13.3, we obtain the main result of this section.

Theorem 13.3. *Let a function $\varphi(x)$ be continuous and bounded in $\mathbb{R}^n$, and a function $f(x,t)$ be continuous and bounded in the layer $\mathbb{R}^n \times (0,T)$ together with its partial derivatives f_{x_i} $(i=1,\dots,n)$. Then the function*

$$u(x,t) = \int_{\mathbb{R}^n} U(x-y,t)\varphi(y)\,dy + \int_0^t ds \int_{\mathbb{R}^n} U(x-y,t-s)f(y,s)\,dy$$

is a classical solution to problem (13.5), (13.6) *and satisfies the inequality*

$$|u(x,t)| \leqslant \sup|\varphi| + T\sup|f|.$$

5 Semigroups of linear operators

14 Uniformly continuous semigroups

Semigroups of linear operators are a subject of modern functional analysis that have important applications in the theory of partial differential equations. To explain this concept and give an initial idea of how the semigroup method works, let us take a look at a number of examples. Unlike Chapter 4, we consider the spaces of complex-valued functions in this chapter.

Example 14.1. Consider the ordinary differential equation

$$\frac{dy}{dt} = ay \tag{14.1}$$

with the initial condition

$$y(0) = y_0, \tag{14.2}$$

where $a \in \mathbb{R}$.

The characteristic equation has the form $\lambda = a$, and the general solution to equation (14.1) is

$$y(t) = ce^{ta},$$

where c is an arbitrary constant. Substituting this solution into the initial condition (14.2), we obtain $c = y_0$. Therefore,

$$y(t) = e^{ta}y_0.$$

Denote the operator of multiplication by the exponent e^{ta} by $T(t)$. Then we write $y(t)$ in the form

$$y(t) = T(t)y_0.$$

Well-known properties of the exponential function,

$$e^{(t+s)a} = e^{ta}e^{sa} \quad \text{and} \quad e^0 = 1,$$

can be rewritten in the form of relations

$$T(t+s) = T(t)T(s) \tag{14.3}$$

and

$$T(0) = 1. \tag{14.4}$$

https://doi.org/10.1515/9783112229637-005

Note also that

$$T(t) \to 1 \quad \text{as } t \to 0. \tag{14.5}$$

Properties (14.3)–(14.5) are used to define a semigroup.

Example 14.2. Consider the system of ordinary differential equations

$$\frac{dy}{dt} = \mathbf{A}y \tag{14.6}$$

with the initial condition

$$y(0) = y_0, \tag{14.7}$$

where $y = y(t)$ is an unknown vector-function, $\mathbf{A}$ is a given matrix of order $n \times n$ with real entries.

The general solution of system (14.6) is represented as

$$y(t) = e^{t\mathbf{A}}C, \tag{14.8}$$

where C is an arbitrary n-dimensional vector and $e^{t\mathbf{A}}$ is the matrix exponential function,

$$e^{t\mathbf{A}} = E + t\mathbf{A} + \frac{t^2\mathbf{A}^2}{2!} + \cdots + \frac{t^k\mathbf{A}^k}{k!} + \cdots.$$

Initial condition (14.7) gives $C = y_0$. Denoting $T(t) = e^{t\mathbf{A}}$, we write

$$y(t) = T(t)y_0. \tag{14.9}$$

The operators $T(t)$ still satisfy the semigroup property, $T(0) = E$ and, in addition, $T(t) \to E$ as $t \to 0$, where E is the identity matrix.

Example 14.3. Consider the heat equation

$$u_t = \Delta u \quad (x \in Q,\ 0 < t < T) \tag{14.10}$$

with the initial condition

$$u|_{t=0} = \varphi(x) \quad (x \in Q). \tag{14.11}$$

Here, $Q \subset \mathbb{R}^n$ is a bounded domain or the entire space $\mathbb{R}^n$.

In the first case, it is also necessary to specify boundary conditions on the cylindrical surface $\partial Q \times (0, T)$. Such a problem was studied in Section 11 by Fourier series with respect to eigenfunctions of the corresponding elliptic operator. In the second case, problem (14.10), (14.11) is the Cauchy problem for the heat equation. On the other hand, both of these problems are studied in a unified way by methods of semigroup theory. Using this approach, we can represent the solution to problem (14.10), (14.11) in the form

$$u(x,t) = T(t)\varphi(x),$$

where $T(t)$ is the semigroup generated by the operator Δ. However, the Laplace operator is unbounded in the corresponding function space. Therefore, we can write $T(t) = e^{\Delta t}$ only formally. This chapter is an introduction to the theory of semigroups of linear operators and its application to problem (14.10), (14.11) and some of its generalizations.

Definition 14.1. Let B be a Banach space. A one-parameter family $\{T(t) : B \to B,\ 0 \leqslant t < \infty\}$ of bounded linear operators in B is called a *semigroup of bounded linear operators* in B if

$$T(t+s) = T(t)T(s) \quad \text{for all } t, s \geqslant 0 \tag{14.12}$$

(*semigroup property*) and

$$T(0) = I, \tag{14.13}$$

where I is the identity operator in B.

Definition 14.2. The linear operator $A\colon B \supset \mathcal{D}(A) \to B$ with the domain

$$\mathcal{D}(A) = \left\{x \in B\colon \text{the limit } \lim_{t\downarrow 0} \frac{T(t)x - x}{t} \text{ exists in } B\right\}, \tag{14.14}$$

acting by the formula

$$Ax = \lim_{t\downarrow 0} \frac{T(t)x - x}{t} = \frac{d^+T(t)x}{dt} \quad (x \in \mathcal{D}(A)), \tag{14.15}$$

is called the *generator* of the semigroup $T(t)$.

Definition 14.3. A semigroup of bounded linear operators $T(t)\colon B \to B$, $t \geqslant 0$, is said to be *uniformly continuous* if

$$\lim_{t\downarrow 0}\|T(t) - I\| = 0. \tag{14.16}$$

Theorem 14.1. *Let $T(t) : B \to B$, $t \geqslant 0$, be a uniformly continuous semigroup of bounded linear operators. Then*

$$\lim_{s\to t}\|T(s) - T(t)\| = 0 \tag{14.17}$$

for all $t \geqslant 0$.

Proof. 1. Take $s = t + h$, where $0 \leqslant t$, $0 < h$. Based on semigroup property (14.12), we write

$$\|T(t+h) - T(t)\| = \|T(t)T(h) - T(t)\| \leqslant \|T(t)\|\|T(h) - I\|. \tag{14.18}$$

From (14.18) and (14.16), we obtain

$$\lim_{s \downarrow t}\|T(s) - T(t)\| = 0. \tag{14.19}$$

2. Let $s = t - h$, where $0 < h < t$, $0 < t$. Again, property (14.12) allows us to write

$$\|T(t-h) - T(t)\| \leqslant \|T(t-h)\| \cdot \|I - T(h)\|, \tag{14.20}$$

$$T(t) = T(t-h)T(h). \tag{14.21}$$

By (14.16), there exists a number h_0, $0 < h_0 < t$ such that $\|T(h) - I\| < 1/2$ for $0 < h < h_0$. Theorem 2.6 and the equality $T(h) = I - (I - T(h))$ imply the existence of a bounded inverse operator $T^{-1}(h)$ with norm $\|T^{-1}(h)\| < 2$ for all $0 < h < h_0$. Therefore, (14.21) means that

$$\|T(t-h)\| \leqslant \|T(t)\| \cdot \|T^{-1}(h)\| \leqslant 2\|T(t)\| \quad \text{for } 0 < h < h_0. \tag{14.22}$$

Now (14.22), (14.20) and (14.16) give

$$\lim_{s \uparrow t}\|T(s) - T(t)\| = 0. \tag{14.23}$$

Obtained relations (14.19) and (14.23) prove (14.17). □

Theorem 14.1 means that a uniformly continuous semigroup is continuous with respect to the operator norm at any point $t \geqslant 0$.

Theorem 14.2. *Let B be a Banach space. A linear operator $A\colon B \supset \mathcal{D}(A) \to B$ is the generator of a uniformly continuous semigroup if and only if A is a bounded operator in B.*

Proof. 1. Let a linear operator $A : B \to B$ be bounded. Set

$$e^{tA} = \sum_{k=0}^{\infty} \frac{(tA)^k}{k!}. \tag{14.24}$$

The series on the right-hand side of (14.24) converges with respect to the operator norm and defines a bounded operator $T(t) = e^{tA}$ for all t. Obviously, $T(0) = I$.

Since the series

$$\sum_{m=0}^{\infty}\sum_{k=0}^{\infty} \frac{t^k A^k}{k!}\,\frac{s^m A^m}{m!}$$

also converges with respect to the operator norm, changing the order of summation in the expression for $T(t)T(s)$ and using the Newton binomial formula, we obtain

$$T(t)T(s) = \sum_{k=0}^{\infty} \frac{t^k A^k}{k!} \sum_{m=0}^{\infty} \frac{s^m A^m}{m!} = \sum_{n=0}^{\infty} \sum_{i=0}^{n} \frac{t^i A^i}{i!} \frac{s^{n-i} A^{n-i}}{(n-i)!} =$$

$$= \sum_{n=0}^{\infty} \frac{A^n}{n!} \sum_{i=0}^{n} \frac{n!}{i!(n-i)!} t^i s^{n-i} = \sum_{n=0}^{\infty} \frac{(t+s)^n A^n}{n!} = T(t+s).$$

The semigroup property is satisfied.

Let us prove that the semigroup $T(t)$ is uniformly continuous. Indeed,

$$\|T(t) - I\| \leqslant \sum_{k=1}^{\infty} \frac{t^k \|A\|^k}{k!} \leqslant \sum_{k=0}^{\infty} \frac{t^{k+1} \|A\|^{k+1}}{k!} = t\|A\| e^{t\|A\|} \to 0 \quad \text{as } t \downarrow 0.$$

Finally, we have

$$\left\| \frac{T(t) - I}{t} - A \right\| \leqslant \sum_{k=2}^{\infty} \frac{t^{k-1} \|A\|^k}{k!} \leqslant \sum_{k=0}^{\infty} \frac{t^{k+1} \|A\|^{k+2}}{k!} = t\|A\|^2 e^{t\|A\|} \to 0$$
$$\text{as } t \downarrow 0.$$

This means that the operator A is the generator of the uniformly continuous semigroup $T(t): B \to B$.

2. Let $T(t): B \to B, t \geqslant 0$, be a uniformly continuous semigroup of bounded linear operators. Therefore, by Theorem 14.1, the function $\|T(t)\|$ is continuous in t on a segment $[0, \rho]$. Then, as is shown in Section 2, the operator function $T(t)$ is Bochner integrable on $[0, \rho]$. Similar arguments can be used throughout the rest of this chapter to justify the existence of the Bochner integral. Moreover, by (14.16) there exists a number $\rho > 0$ such that $\|T(t) - I\| < 1$ for $t \in [0, \rho]$. Therefore, using inequality (2.2), we obtain

$$\left\| \frac{1}{\rho} \int_0^{\rho} T(t)\,dt - I \right\| = \left\| \frac{1}{\rho} \int_0^{\rho} (T(t) - I)\,dt \right\| \leqslant \frac{1}{\rho} \int_0^{\rho} \|T(t) - I\|\,dt <$$
$$< \frac{1}{\rho} \int_0^{\rho} dt = 1.$$

It follows now from Theorem 2.6 that the operator

$$\frac{1}{\rho} \int_0^{\rho} T(t)\,dt = I - \left(I - \frac{1}{\rho} \int_0^{\rho} T(t)\,dt \right)$$

is invertible in B. Therefore, the operator $\int_0^{\rho} T(t)\,dt$ itself has a bounded inverse.

Let us prove that the operator

$$A = (T(\rho) - I)\left(\int_0^\rho T(s)\,ds\right)^{-1}$$

is the generator of the semigroup $T(t)$. It is clear that

$$\left\|\frac{T(h) - I}{h} - (T(\rho) - I)\left(\int_0^\rho T(s)\,ds\right)^{-1}\right\| \leqslant$$

$$\leqslant \left\|\frac{T(h) - I}{h}\int_0^\rho T(s)\,ds - T(\rho) + I\right\| \cdot \left\|\left(\int_0^\rho T(s)\,ds\right)^{-1}\right\|. \tag{14.25}$$

Based on the semigroup property, as well as on the properties of integrals of operator-functions (see Section 2), we have

$$\left\|\frac{T(h) - I}{h}\int_0^\rho T(s)\,ds - T(\rho) + I\right\| =$$

$$= \left\|\frac{1}{h}\int_0^\rho (T(s+h) - T(s))\,ds - T(\rho) + I\right\| =$$

$$= \left\|\frac{1}{h}\int_\rho^{\rho+h} (T(s) - T(\rho))\,ds + \frac{1}{h}\int_0^h (I - T(s))\,ds\right\| \leqslant$$

$$\leqslant \sup_{\rho \leqslant s \leqslant \rho+h} \|T(s) - T(\rho)\| + \sup_{0 \leqslant s \leqslant h} \|T(s) - I\|. \tag{14.26}$$

By Theorem 14.1, the expression on the right-hand side of (14.26) tends to zero as $h \downarrow 0$; hence

$$A = \lim_{h \downarrow 0} \frac{T(h) - I}{h} = (T(\rho) - I)\left(\int_0^\rho T(s)\,dt\right)^{-1},$$

where the convergence takes place with respect to the operator norm. Obviously, $\mathcal{D}(A) = B$. It is shown that if a semigroup $T(t): B \to B$, $t \geqslant 0$, is uniformly continuous, then its generator A is a bounded linear operator in B. □

The statement below follows from the proof of Theorem 14.2.

Corollary 14.1. *Let a bounded linear operator $A : B \to B$ be the generator of a uniformly continuous semigroup $T(t): B \to B$ $(t \geqslant 0)$. Then*

$$\lim_{h \downarrow 0} \left\|\frac{T(h) - I}{h} - A\right\| = 0.$$

Theorem 14.3. *Let $T(t), S(t) : B \to B$ $(t \geqslant 0)$ be two uniformly continuous semigroups of bounded linear operators. If a bounded linear operator $A : B \to B$ is the generator of both semigroups $T(t)$ and $S(t)$, then these semigroups coincide, i. e., $T(t) = S(t)$ for $t \geqslant 0$. In other words, every uniformly continuous semigroup is uniquely determined by its generator.*

Proof. Pick an arbitrary number $t_0 > 0$. Let us prove that $T(t) = S(t)$ for $0 \leqslant t \leqslant t_0$. By Theorem 14.1, the functions $t \mapsto \|T(t)\|$ and $t \mapsto \|S(t)\|$ are continuous on $[0, T]$. Therefore,

$$\|T(t)\|\|S(s)\| \leqslant C \quad \text{for } 0 \leqslant s,\ t \leqslant t_0 \tag{14.27}$$

for some constant $C > 0$. Corollary 14.1 shows that for any $\varepsilon > 0$ there exists $\delta > 0$ such that

$$h^{-1}\|T(h) - S(h)\| \leqslant \left\|\frac{T(h) - I}{h} - A\right\| + \left\|\frac{S(h) - I}{h} - A\right\| < \frac{\varepsilon}{t_0 C} \quad \text{for } 0 < h < \delta. \tag{14.28}$$

Let $0 \leqslant t \leqslant t_0$. We choose $n \geqslant 1$ such that $t_0/n < \delta$. From the semigroup property and inequalities (14.27), (14.28), we obtain

$$\begin{aligned}
\|T(t) - S(t)\| &= \left\|T\left(n\frac{t}{n}\right) - S\left(n\frac{t}{n}\right)\right\| \leqslant \\
&\leqslant \sum_{k=0}^{n-1}\left\|T\left((n-k)\frac{t}{n}\right)S\left(\frac{kt}{n}\right) - T\left((n-k-1)\frac{t}{n}\right)S\left(\frac{(k+1)t}{n}\right)\right\| \leqslant \\
&\leqslant \sum_{k=0}^{n-1}\left\|T\left((n-k-1)\frac{t}{n}\right)\right\| \cdot \left\|T\left(\frac{t}{n}\right) - S\left(\frac{t}{n}\right)\right\| \cdot \left\|S\left(\frac{kt}{n}\right)\right\| \leqslant \\
&\leqslant Cn\frac{\varepsilon}{t_0 C}\frac{t}{n} \leqslant \varepsilon.
\end{aligned}$$

Since $\varepsilon > 0$ is arbitrary, $T(t) = S(t)$ for $0 \leqslant t \leqslant t_0$. □

Corollary 14.2. *Let $T(t) : B \to B$ $(t \geqslant 0)$ be a uniformly continuous semigroup of bounded linear operators. Then the following statements are true:*

(a) *There exists a unique linear operator $A : B \to B$ such that $T(t) = e^{tA}$.*
(b) *There exists a constant $\omega \geqslant 0$ such that $\|T(t)\| \leqslant e^{\omega t}$.*
(c) *The operator-function $[0, \infty) \ni t \mapsto T(t)$ is differentiable with respect to the operator norm and*

$$\frac{dT(t)}{dt} = AT(t) = T(t)A \quad (t \geqslant 0). \tag{14.29}$$

Proof. 1. Let us prove statement (a). It follows from Theorem 14.2 that the generator of the semigroup $T(t) : B \to B$, $t \geqslant 0$, is a bounded linear operator $A : B \to B$. At the same time, as can be seen from the proof of Theorem 14.2, the operator A is the generator of the semigroup e^{tA}. By Theorem 14.3, we obtain $T(t) = e^{At}$. The uniqueness of A is obvious.

2. Statement (b) obviously follows from (a).

3. Let us move on to statement (c). It follows from the semigroup property that

$$\frac{T(t+h)-T(t)}{h} = \frac{T(h)-I}{h}T(t) = T(t)\frac{T(h)-I}{h} \tag{14.30}$$

(for any fixed $t \geqslant 0$ and all $h > 0$) and

$$\frac{T(t-h)-T(t)}{-h} = \frac{T(h)-I}{h}T(t-h) = T(t-h)\frac{T(h)-I}{h} \tag{14.31}$$

(for any fixed $t > 0$ and all $0 < h < t$).

Passing to the limit as $h \downarrow 0$ in combination with Corollary 14.1 and Theorem 14.1, we have

$$\frac{d^+T(t)}{dt} = AT(t) = T(t)A \quad (t \geqslant 0), \tag{14.32}$$

$$\frac{d^-T(t)}{dt} = AT(t) = T(t)A \quad (t > 0). \tag{14.33}$$

From (14.32) and (14.33), we obtain (14.29). □

Let us consider examples related to uniformly continuous semigroups.

Example 14.4. Let a linear transformation $A: \mathbb{R}^2 \to \mathbb{R}^2$ have the matrix $\mathbf{A} = \left(\begin{smallmatrix} 2 & 9 \\ 1 & 2 \end{smallmatrix}\right)$ in the basis $e_1 = \left(\begin{smallmatrix} 1 \\ 0 \end{smallmatrix}\right)$, $e_2 = \left(\begin{smallmatrix} 0 \\ 1 \end{smallmatrix}\right)$. By Theorem 14.2 and Corollary 14.2, the operator $A: \mathbb{R}^2 \to \mathbb{R}^2$ is the generator of the uniformly continuous semigroup $T(t) = e^{tA}: \mathbb{R}^2 \to \mathbb{R}^2$, $t \geqslant 0$. Let us obtain an explicit representation for the matrix exponential e^{tA}. The characteristic equation has the form

$$|\mathbf{A} - \lambda\mathbf{E}| = \begin{vmatrix} 2-\lambda & 9 \\ 1 & 2-\lambda \end{vmatrix} = \lambda^2 - 4\lambda - 5 = 0,$$

thus the eigenvalues of the matrix $\mathbf{A}$ are $\lambda_1 = 5, \lambda_2 = -1$. Find corresponding eigenvectors for $\mathbf{A}$. The equation $\mathbf{A}e_1' = \lambda_1 e_1'$ is represented as the system

$$\begin{aligned} -3c_{11} + 9c_{21} &= 0, \\ c_{11} - 3c_{21} &= 0, \end{aligned}$$

where $e_1' = \left(\begin{smallmatrix} c_{11} \\ c_{21} \end{smallmatrix}\right)$ is an eigenvector of $\mathbf{A}$ corresponding to the eigenvalue λ_1. By solving this system, we obtain $e_1' = \left(\begin{smallmatrix} 3 \\ 1 \end{smallmatrix}\right)$.

The equality $\mathbf{A}e_2' = \lambda_2 e_2'$ is equivalent to the system

$$\begin{aligned} 3c_{12} + 9c_{22} &= 0, \\ c_{12} + 3c_{22} &= 0, \end{aligned}$$

where $e_2' = \binom{c_{12}}{c_{22}}$ is an eigenvector of $\mathbf{A}$ corresponding to the eigenvalue λ_2. We get $e_2' = \binom{3}{-1}$.

Next, we compose the transition matrix from the basis e_1, e_2 to the basis e_1', e_2':

$$\mathbf{T} = \begin{pmatrix} 3 & 3 \\ 1 & -1 \end{pmatrix}.$$

In the basis e_1', e_2' of the eigenvectors for $\mathbf{A}$, a matrix of the linear transformation matrix A has the form

$$\mathbf{A}' = \begin{pmatrix} 5 & 0 \\ 0 & -1 \end{pmatrix},$$

and the well-known relation $\mathbf{A} = \mathbf{T}\mathbf{A}'\mathbf{T}^{-1}$ gives the connection between the two matrices of this transformation. The following calculations are obvious:

$$\begin{aligned} e^{t\mathbf{A}} &= \mathbf{T}e^{t\mathbf{A}'}\mathbf{T}^{-1} = \begin{pmatrix} 3 & 3 \\ 1 & -1 \end{pmatrix} \begin{pmatrix} e^{5t} & 0 \\ 0 & e^{-t} \end{pmatrix} \begin{pmatrix} \frac{1}{6} & \frac{1}{2} \\ \frac{1}{6} & -\frac{1}{2} \end{pmatrix} = \\ &= \begin{pmatrix} \frac{1}{2}e^{5t} + \frac{1}{2}e^{-t} & \frac{3}{2}e^{5t} - \frac{3}{2}e^{-t} \\ \frac{1}{6}e^{5t} - \frac{1}{6}e^{-t} & \frac{1}{2}e^{5t} + \frac{1}{2}e^{-t} \end{pmatrix}. \end{aligned}$$

Example 14.5. Let a linear transformation matrix $A\colon \mathbb{C}^2 \to \mathbb{C}^2$ have the form

$$\mathbf{A} = \begin{pmatrix} 1 & 2 \\ -2 & 1 \end{pmatrix}.$$

Here, we also write the corresponding uniformly continuous semigroup as the matrix exponential $T(t) = e^{tA}\colon \mathbb{C}^2 \to \mathbb{C}^2$, $t \geqslant 0$. Let us get the explicit form for $e^{t\mathbf{A}}$. We calculate the eigenvalues:

$$|\mathbf{A} - \lambda\mathbf{E}| = \begin{vmatrix} 1-\lambda & 2 \\ -2 & 1-\lambda \end{vmatrix} = (\lambda - 1)^2 + 4 = 0,$$

whence $\lambda_1 = 1 + 2i$, $\lambda_2 = 1 - 2i$.

Next, we find the eigenvectors of the matrix $\mathbf{A}$. For the eigenvector $e_1' = \binom{c_{11}}{c_{21}}$ corresponding to the eigenvalue λ_1, $Ae_1' = \lambda_1 e_1'$, we have the system

$$\begin{aligned} -2ic_{11} + 2c_{21} &= 0, \\ -2c_{11} - 2ic_{21} &= 0, \end{aligned}$$

whence $e_1' = \binom{1}{i}$.

The eigenvector $e_2' = \binom{c_{12}}{c_{22}}$ corresponding to the eigenvalue λ_2, $Ae_2' = \lambda_2 e_2'$, is obtained from the system

$$2ic_{12} + 2c_{22} = 0,$$
$$-2c_{12} + 2ic_{22} = 0.$$

We obtain $e_2' = \binom{1}{-i}$.

We write down the transition matrix from the basis e_1, e_2 to the basis e_1', e_2':

$$\mathbf{T} = \begin{pmatrix} 1 & 1 \\ i & -i \end{pmatrix}.$$

In the basis e_1', e_2' of eigenvectors for $\mathbf{A}$, the matrix of the linear transformation is

$$\mathbf{A}' = \begin{pmatrix} 1+2i & 0 \\ 0 & 1-2i \end{pmatrix}.$$

Using the relation $\mathbf{A} = \mathbf{T}\mathbf{A}'\mathbf{T}^{-1}$, we obtain

$$e^{t\mathbf{A}} = \mathbf{T}e^{t\mathbf{A}'}\mathbf{T}^{-1} = \begin{pmatrix} 1 & 1 \\ i & -i \end{pmatrix} \begin{pmatrix} e^{(1+2i)t} & 0 \\ 0 & e^{(1-2i)t} \end{pmatrix} \begin{pmatrix} \frac{1}{2} & -\frac{i}{2} \\ \frac{1}{2} & \frac{i}{2} \end{pmatrix} =$$
$$= \begin{pmatrix} e^t \cos 2t & e^t \sin 2t \\ -e^t \sin 2t & e^t \cos 2t \end{pmatrix}.$$

15 Strongly continuous semigroups

Let B be a Banach space.

Definition 15.1. A semigroup of bounded linear operators $T(t): B \to B$, $t \geqslant 0$ is said to be *strongly continuous* if

$$\lim_{t\downarrow 0} \|T(t)x - x\| = 0 \quad \text{for all } x \in B. \tag{15.1}$$

A strongly continuous semigroup of bounded linear operators is also called a C_0-*semigroup.*

Theorem 15.1. *Let $T(t): B \to B$, $t \geqslant 0$ be a C_0-semigroup. Then there exist constants $\omega \geqslant 0$ and $M \geqslant 1$ such that*

$$\|T(t)\| \leqslant Me^{\omega t} \quad \textit{for } t \geqslant 0. \tag{15.2}$$

Proof. 1. Let us prove the existence of a number $t_0 > 0$ such that the norm $\|T(t)\|$ is bounded on the interval $0 \leqslant t \leqslant t_0$. Assume the opposite: there exists a sequence $\{t_n\}$ such that $t_n \geqslant 0$, $\lim_{n\to\infty} t_n = 0$ and $\|T(t_n)\| \geqslant n$. But then by the Banach–Steinhaus theorem, the number sequence $\|T(t_n)x\|$ is unbounded for some vector $x \in B$. This contradicts (15.1). This means that $\|T(t)\| \leqslant M$ for $0 \leqslant t \leqslant t_0$. Since $\|T(0)\| = 1$, we have $M \geqslant 1$.

2. Denote $\omega = t_0^{-1} \ln M$. It is clear that $\omega \geqslant 0$. Any number $t \geqslant 0$ can be represented as $t = kt_0 + \delta$, where $k \in \mathbb{Z}$, $k \geqslant 0$ and $0 \leqslant \delta < t_0$. Then, using the semigroup property, we obtain

$$\|T(t)\| = \|T(t_0)^k T(\delta)\| \leqslant M^{k+1} \leqslant MM^{t/t_0} = Me^{\omega t}.$$ □

Theorem 15.2. *Let $T(t): B \to B$, $t \geqslant 0$, be a C_0-semigroup. Then the vector function $[0, \infty) \ni t \mapsto T(t)x \in B$ is continuous for each vector $x \in B$.*

Proof. The semigroup property and Theorem 15.1 imply the estimate

$$\|T(t+h)x - T(t)x\| \leqslant \|T(t)\| \|T(h)x - x\| \leqslant Me^{\omega t} \|T(h)x - x\| \tag{15.3}$$

for $x \in B$, $t \geqslant 0$ and $h > 0$, as well as the estimate

$$\begin{aligned} \|T(t-h)x - T(t)x\| &\leqslant \|T(t-h)\| \cdot \|x - T(h)x\| \leqslant \\ &\leqslant Me^{\omega t} \|x - T(h)x\| \end{aligned} \tag{15.4}$$

for $x \in B$, $t > 0$ and $0 < h < t$.

Relations (15.3), (15.4) and (15.1) immediately imply the continuity of the vector function $[0, \infty) \ni t \mapsto T(t)x \in B$ for all $x \in B$. □

Theorem 15.3. *Let $T(t): B \to B$, $t \geqslant 0$ be a C_0-semigroup, and $A: B \supset \mathcal{D}(A) \to B$ be its generator. Then the following statements are true:*

(a) *If $x \in B$, then*

$$\lim_{h \to 0} \frac{1}{h} \int_t^{t+h} T(s)x\, ds = T(t)x. \tag{15.5}$$

(b) *If $x \in B$, then $\int_0^t T(s)x\, ds \in \mathcal{D}(A)$ and*

$$A \int_0^t T(s)x\, ds = T(t)x - x. \tag{15.6}$$

(c) *If $x \in \mathcal{D}(A)$, then $T(t)x \in \mathcal{D}(A)$ and*

$$\frac{d}{dt} T(t)x = AT(t)x = T(t)Ax. \tag{15.7}$$

(d) *If $x \in \mathcal{D}(A)$, then*

$$T(t)x - T(s)x = \int_s^t T(\tau)Ax\, d\tau = \int_s^t AT(\tau)x\, d\tau. \tag{15.8}$$

Proof. 1. First, we prove statement (a). Let $x \in B$. By Theorem 15.2, for any $t \geqslant 0$ and $\varepsilon > 0$ there exists a number $h \neq 0$ such that $\|T(s)x - T(t)x\| < \varepsilon$ for $|s - t| < |h|$. Using the properties of the integral of the vector function, we have

$$\left\| \frac{1}{h} \int_t^{t+h} T(s)x\,ds - T(t)x \right\| = \left\| \frac{1}{h} \int_t^{t+h} (T(s)x - T(t)x)\,ds \right\| \leqslant$$
$$\leqslant \left| \frac{1}{h} \int_t^{t+h} \|T(s)x - T(t)x\|\,ds \right| < \varepsilon.$$

2. Let $x \in B$ and $h > 0$. Based on the semigroup property, we can write

$$\frac{T(h) - I}{h} \int_0^t T(s)x\,ds = \frac{1}{h} \int_0^t (T(s+h)x - T(s)x)\,ds =$$
$$= \frac{1}{h} \int_t^{t+h} T(s)x\,ds - \frac{1}{h} \int_0^h T(s)x\,ds. \tag{15.9}$$

In view of statement (a), the expression on the right-hand side tends to $T(t)x - x$ as $h \downarrow 0$. Therefore, the expression on the left has a limit. By definition of $\mathcal{D}(A)$ this means that $\int_0^t T(s)x\,ds \in \mathcal{D}(A)$, and the limit itself is equal to $A \int_0^t T(s)x\,ds$. This proves (b).

3. Let us move on to (c). Now suppose that $x \in \mathcal{D}(A)$ and $h > 0$. For all $t \geqslant 0$, the following relation holds:

$$\frac{T(t+h) - T(t)}{h} x = \frac{T(h) - I}{h} T(t)x = T(t) \frac{T(h) - I}{h} x \to$$
$$\to T(t)Ax \quad \text{as } h \downarrow 0.$$

Therefore, $T(t)x \in \mathcal{D}(A)$ and

$$\frac{d^+}{dt} T(t)x = AT(t)x = T(t)Ax,$$

i. e., the right derivative of the function $T(t)x$ exists and is equal to $T(t)Ax$. It remains to show that for $t > 0$ the left derivative of $T(t)x$ also exists and is equal to $T(t)Ax$. Obviously,

$$\left\| \frac{T(t)x - T(t-h)x}{h} - T(t)Ax \right\| \leqslant$$
$$\leqslant \|T(t-h)\| \left\| \frac{T(h)x - x}{h} - Ax \right\| + \|T(t-h)Ax - T(t)Ax\|. \tag{15.10}$$

By Theorem 15.1, we have $\|T(t-h)\| \leqslant Me^{\omega t}$. Therefore, the first term on the right-hand side of (15.10) tends to zero as $h \to 0$. It follows from Theorem 15.2 that the second term on the right-hand side of (15.10) also tends to zero as $h \to 0$. Therefore, the left derivative of the function $T(t)x$ exists and the following equality holds:

$$\frac{d^- T(t)x}{dt} = T(t)Ax.$$

4. Integrating (15.7) from s to t, we obtain (15.8). □

Let us recall a well-known definition.

Definition 15.2. A linear operator $A: B \supset \mathcal{D}(A) \to B$ is called *closed* if the assumptions $x_k \in \mathcal{D}(A)$, $x_k \to x$ and $Ax_k \to y$ as $k \to \infty$ imply $x \in \mathcal{D}(A)$ and $Ax = y$.

Theorem 15.4. *Let a linear operator $A: B \supset \mathcal{D}(A) \to B$ be the generator of a C_0-semigroup $T(t): B \to B$, $t \geqslant 0$. Then the domain of $\mathcal{D}(A)$ is dense in B and the operator A is closed.*

Proof. 1. For each $x \in B$, set $x_t = \frac{1}{t}\int_0^t T(s)x\,ds$. From statement (b) of Theorem 15.3 it follows that $x_t \in \mathcal{D}(A)$ for $t > 0$. On the other hand, in view of point (a) of the same theorem, $x_t \to x$ as $t \downarrow 0$. Therefore, the closure of $\mathcal{D}(A)$ coincides with B.

2. Let us prove that the operator A is closed. Suppose that $x_k \in \mathcal{D}(A)$, $x_k \to x$ and $Ax_k \to y$ as $k \to \infty$. From statement (d) of Theorem 15.3, it follows that

$$T(t)x_k - x_k = \int_0^t T(s)Ax_k\,ds. \tag{15.11}$$

By Theorem 15.1, the expression under the integral on the right-hand side of (15.11) converges to $T(s)y$ uniformly on bounded intervals. Therefore, passing to the limit in (15.11) as $k \to \infty$, we get

$$T(t)x - x = \int_0^t T(s)y\,ds. \tag{15.12}$$

Divide both sides of (15.12) by $t > 0$. Passing to the limit as $t \downarrow 0$ and using statement (a) of Theorem 15.3, we conclude that $x \in \mathcal{D}(A)$ and $Ax = y$. □

Theorem 15.5. *Let $T(t)$, $S(t): B \to B$, $t \geqslant 0$, be C_0-semigroups. If a linear operator $A: B \supset \mathcal{D}(A) \to B$ is the generator of both semigroups $T(t)$ and $S(t)$, then $T(t) = S(t)$ for $t \geqslant 0$. In other words, a strongly continuous semigroup is uniquely determined by its generator.*

Proof. Let $x \in \mathcal{D}(A)$ and $0 \leqslant s \leqslant t$. Then, by (15.7) we have

$$\frac{d}{ds}\big(T(t-s)S(s)x\big) = -T(t-s)AS(s)x + T(t-s)AS(s)x = 0.$$

Therefore, the vector function $T(t-s)S(s)x$ does not depend on s, which allows us to write

$$T(t)x = T(t-s)S(s)x\big|_{s=0} = T(t-s)S(s)x\big|_{s=t} = S(t)x.$$

We have proven that $T(t)x = S(t)x$ for all $x \in \mathcal{D}(A)$ for $t \geqslant 0$. Since the domain of $\mathcal{D}(A)$ is dense in B and the operators $T(t)$ and $S(t)$ are bounded, we conclude that $T(t)x = S(t)x$ for all $x \in X$ for $t \geqslant 0$. Therefore, the operator A generates a unique semigroup. □

Remark 15.1. It is clear that a uniformly continuous semigroup is strongly continuous. The converse is not true, as the example below shows.

Example 15.1. Consider a family of operators $T(t)\colon L_2(0,1) \to L_2(0,1)$, $t \geqslant 0$, given by the formula

$$T(t)\varphi(x) = \begin{cases} 0, & 0 < x \leqslant t, \\ \varphi(x-t), & t < x < 1. \end{cases}$$

It is clear that the operators $T(t)$ are linear. We can easily see their boundedness:

$$\|T(t)\varphi\|^2_{L_2(0,1)} = \int_t^1 |\varphi(x-t)|^2 dx \leqslant \int_0^1 |\varphi(x)|^2 dx = \|\varphi\|^2_{L_2(0,1)}.$$

The family $T(t)\colon L_2(0,1) \to L_2(0,1)$, $t \geqslant 0$, is a semigroup. By definition, $T(0)\varphi(x) = \varphi(x)$ for $0 < x < 1$, i. e., $T(0) = I$. Moreover,

$$T(s)T(t)\varphi(x) = \begin{cases} 0, & 0 < x \leqslant t+s, \\ \varphi(x-t-s), & t+s < x < 1, \end{cases}$$

i. e., $T(s)T(t) = T(s+t)$, $t, s \geqslant 0$.

The semigroup $T(t)\colon L_2(0,1) \to L_2(0,1)$ is strongly continuous. Indeed, by Theorem 4.2,

$$\|T(t)\varphi - \varphi\|^2_{L_2(0,1)} = \int_0^1 |\varphi(x-t) - \varphi(x)|^2 dx \to 0 \quad \text{as } t \downarrow 0.$$

Here, we took into account that $\varphi(x-t) = 0$ for $0 < x \leqslant t$.

Let us find the generator of the semigroup $T(t)$. Let $\varphi \in \mathcal{D}(A)$, i. e., there exist an element $g \in L_2(0,1)$ such that

$$\lim_{t\downarrow 0} \left\| \frac{T(t)\varphi - \varphi}{t} - g \right\|_{L_2(0,1)} = 0,$$

whence it follows, in particular, that

$$\left\| \frac{T(t)\varphi - \varphi}{t} \right\|_{L_2(0,1)} \leqslant C = \|g\|_{L_2(0,1)} + 1 \quad \text{for } 0 < t < \delta$$

if $\delta > 0$ is small enough.

Let $\Phi(x)$ denote the extension of the function $\varphi(x)$ by zero for $-1 < x < 0$. Then

$$\left\|\frac{\Phi(x-t)-\Phi(x)}{-t}\right\|_{L_2(-1,1)} = \left\|\frac{T(t)\varphi(x)-\varphi(x)}{t}\right\|_{L_2(0,1)} \leqslant C \quad (0 < t < x).$$

Repeating the proof of statement (b) in Theorem 5.5, we make sure that there exists a weak derivative $\Phi' \in L_2(-1,1)$, i. e., $\Phi \in H^1(-1,1)$. The Sobolev embedding theorem guarantees that $\Phi(x)$ coincides a. e. with a $C[-1,1]$-function. Therefore, since $\Phi(x) = 0$ for $x < 0$, we have $\Phi(0) = 0$. Due to the relations $\Phi \in H^1(0,1)$ and $\Phi(x) = \varphi(x)$ for $x \in (0,1)$, we also have $\varphi \in H^1(0,1)$ and $\varphi(0) = 0$. Therefore,

$$\mathcal{D}(A) \subset \{\varphi \in H^1(0,1): \varphi(0) = 0\}.$$

Prove the converse inclusion. Assuming $\varphi \in H^1(0,1)$ and $\varphi(0) = 0$, make sure that $\varphi \in \mathcal{D}(A)$.

Let again $\Phi(x)$ be the extension of the function $\varphi(x)$ by zero to the interval $-1 < x < 0$, then $\Phi \in H^1(-1,1)$. Obviously,

$$\left\|\frac{T(t)\varphi-\varphi}{t} - (-\Phi')\right\|_{L_2(0,1)} = \left\|\frac{\Phi(x-t)-\Phi(x)}{-t} - \Phi'(x)\right\|_{L_2(-1,1)}.$$

Proceeding similar to the proof of statement (a) of Theorem 5.5, we see that the norm on the right-hand side tends to zero as $t \downarrow 0$. Since $\Phi(x) = \varphi(x)$ for $x \in (0,1)$, we deduce that $\varphi \in \mathcal{D}(A)$ and $A\varphi = -\varphi'(x)$.

Thus, we have shown that $\mathcal{D}(A) = \{\varphi \in H^1(0,1): \varphi(0) = 0\}$ and $A\varphi = -\varphi'(x)$ $(x \in (0,1))$.

The operator $A: \mathcal{D}(A) \subset L_2(0,1) \to L_2(0,1)$ is not bounded. Taking the sequence $f_k(x) = \sin \pi k x$, we see that $\|f_k\|_{L_2(0,1)} = \frac{1}{\sqrt{2}}$ but $Af_k(x) = -\pi k \cos \pi k x$ and $\|Af_k\|_{L_2(0,1)} = \frac{\pi k}{\sqrt{2}} \to \infty$. In view of Theorem 14.2, we can conclude that the semigroup of bounded linear operators $T(t): L_2(0,1) \to L_2(0,1)$ is not uniformly continuous.

Example 15.2. Consider the semigroup of operators $T(t): L_2(\mathbb{R}) \to L_2(\mathbb{R})$ defined by the formula $T(t)f(x) = f(x+t)$, $t \geqslant 0$. It is easy to show that the generator of this semigroup is the differentiation operator $Af = f'$ with domain $\mathcal{D}(A) = H^1(\mathbb{R})$.

16 Hille–Yosida theorem

Let $T(t): B \to B$, $t \geqslant 0$, be a C_0-semigroup. By Theorem 15.1, there exist constants $\omega \geqslant 0$ and $M \geqslant 1$ such that $\|T(t)\| \leqslant Me^{\omega t}$ for $t \geqslant 0$.

Definition 16.1. A strongly continuous semigroup of bounded linear operators $T(t): B \to B$, $t \geqslant 0$ is called a *uniformly bounded C_0-semigroup* if $\|T(t)\| \leqslant M$, $t \geqslant 0$. If $\|T(t)\| \leqslant 1$, $t \geqslant 0$, then the semigroup is called a *contractive C_0-semigroup*.

In contrast to uniformly continuous semigroups, the generator of an arbitrary C_0-semigroup is, generally speaking, an unbounded operator (see Examples 15.1, 15.2). The following statement contains necessary and sufficient conditions for a linear operator $A: B \supset \mathcal{D}(A) \to B$ to be the generator of a contractive C_0-semigroup. This fundamental result is called the Hille–Yosida theorem.

Let us recall the definition of a resolvent. A complex number λ is called a *regular value* of a linear operator $A: B \supset \mathcal{D}(A) \to B$ if the operator $\lambda I - A$ maps $\mathcal{D}(A)$ one-to-one onto the entire space B and the inverse operator $(\lambda I - A)^{-1} : B \to B$ is bounded. The set of all regular values of the operator A is called the *resolvent set* of the operator A and is denoted by $\rho(A)$. The operator-function $(\lambda I - A)^{-1}$ depending on $\lambda \in \rho(A)$ is called the *resolvent* of the operator A and is denoted by $R(\lambda, A)$.

Theorem 16.1. *A linear operator $A: B \supset \mathcal{D}(A) \to B$ is the generator of a contractive C_0-semigroup $T(t): B \to B$, $t \geqslant 0$, if and only if:*

(a) *A is closed and $\overline{\mathcal{D}(A)} = B$.*

(b) *The resolvent set $\rho(A)$ of the operator A contains the positive real semiaxis $\mathbb{R}_+$, and the inequality*

$$\|R(\lambda, A)\| \leqslant \frac{1}{\lambda} \tag{16.1}$$

holds for all $\lambda > 0$.

Proof of Theorem 16.1 (*necessity*). Let a linear operator $A: B \supset \mathcal{D}(A) \to B$ be the generator of a contractive C_0-semigroup $T(t): B \to B$, $t \geqslant 0$. By Theorem 15.4, the operator A is closed and $\overline{\mathcal{D}(A)} = B$. Let us prove property (b).

For $\lambda > 0$ and $x \in B$, we introduce

$$R(\lambda)x = \int_0^\infty e^{-\lambda t} T(t)x\, dt. \tag{16.2}$$

By Theorem 15.2, the function $[0, \infty) \ni t \mapsto T(t)x \in B$ is continuous. Since the semigroup $T(t): B \to B$, $t \geqslant 0$ is contractive, this function is uniformly bounded on $[0, \infty)$. Therefore, the integral of the norm of the integrand on the right-hand side of (16.2) exists as an improper Riemann integral. Using again the contraction property of the semigroup $T(t)$, we obtain

$$\|R(\lambda)x\| \leqslant \int_0^\infty e^{-\lambda t} \|T(t)x\|\, dt \leqslant \frac{\|x\|}{\lambda}. \tag{16.3}$$

It remains to prove that the operator $R(\lambda)$ is the inverse of the operator $\lambda I - A$. Based on the semigroup property, we write

$$\frac{T(h)-I}{h}R(\lambda)x = \frac{1}{h}\int_0^\infty e^{-\lambda t}(T(t+h)x - T(t)x)\,dt =$$
$$= \frac{e^{\lambda h}-1}{h}\int_0^\infty e^{-\lambda t}T(t)x\,dt - \frac{e^{\lambda h}}{h}\int_0^h e^{-\lambda t}T(t)x\,dt \tag{16.4}$$

for $h > 0$.

By virtue of Theorem 15.3, statement (a), the expression on the right-hand side of (16.4) converges to the vector $\lambda R(\lambda)x - x$ as $h \downarrow 0$. This means that for all $x \in B$ and $\lambda > 0$ the relations $R(\lambda)x \in \mathcal{D}(A)$ and $AR(\lambda)x = (\lambda R(\lambda) - I)x$ are valid, i. e.,

$$(\lambda I - A)R(\lambda)x = x \quad (x \in B). \tag{16.5}$$

On the other hand, statement (c) of Theorem 15.3, together with the closedness of A, allows us to write

$$R(\lambda)Ax = \int_0^\infty e^{-\lambda t}T(t)Ax\,dt = \int_0^\infty e^{-\lambda t}AT(t)x\,dx =$$
$$= A\left(\int_0^\infty e^{-\lambda t}T(t)x\,dt\right) = AR(\lambda)x \quad (x \in \mathcal{D}(A)). \tag{16.6}$$

Equations (16.5) and (16.6) mean that

$$R(\lambda)(\lambda I - A)x = x \quad (x \in \mathcal{D}(A)). \tag{16.7}$$

Therefore, the operator $R(\lambda)$ is defined for all $\lambda > 0$ and the equality $R(\lambda) = (\lambda I - A)^{-1}$ is valid. In addition, estimate (16.1) is satisfied due to inequality (16.3). The necessity of conditions (a) and (b) is proven. □

To prove sufficiency, we need a number of auxiliary results.

Lemma 16.1. *Let an operator* $A: B \supset \mathcal{D}(A) \to B$ *satisfy conditions (a) and (b) of Theorem* 16.1. *Then*

$$\lim_{\lambda\to\infty} \lambda R(\lambda, A)x = x \quad (x \in B). \tag{16.8}$$

Proof. 1. Assume that $x \in \mathcal{D}(A)$. Then, by (16.1), we have

$$\|\lambda R(\lambda, A)x - x\| = \|AR(\lambda, A)x\| = \|R(\lambda, A)Ax\| \leqslant$$
$$\leqslant \frac{1}{\lambda}\|Ax\| \to 0 \quad \text{for } \lambda \to \infty.$$

2. Take $x \in B$. Since the domain $\mathcal{D}(A)$ of the operator A is dense in B, there exists for every $\varepsilon > 0$ an element $x_\varepsilon \in \mathcal{D}(A)$ such that $\|x - x_\varepsilon\| < \varepsilon$. Therefore, it follows from

inequality (16.1) and the first part of the proof that

$$\begin{aligned}\|\lambda R(\lambda, A)x - x\| &= \|\lambda R(\lambda, A)(x - x_\varepsilon) + \lambda R(\lambda, A)x_\varepsilon - x_\varepsilon + x_\varepsilon - x\| \leqslant \\ &\leqslant \|\lambda R(\lambda, A)(x - x_\varepsilon)\| + \|\lambda R(\lambda, A)x_\varepsilon - x_\varepsilon\| + \|x_\varepsilon - x\| < 3\varepsilon\end{aligned}$$

if the number $\lambda > 0$ is large enough. Therefore,

$$\lim_{\lambda\to\infty} \|\lambda R(\lambda, A)x - x\| = 0$$

for all $x \in B$. □

A linear bounded operator $A_\lambda: B \to B$, $\lambda > 0$, defined by the formula

$$A_\lambda = \lambda AR(\lambda, A) = \lambda^2 R(\lambda, A) - \lambda I, \tag{16.9}$$

is called the *Yosida approximation* of the operator A.

Lemma 16.2. *Let an operator $A: B \supset \mathcal{D}(A) \to B$ satisfy conditions (a) and (b) of Theorem* 16.1. *Then*

$$\lim_{\lambda\to\infty} A_\lambda x = Ax \quad \textit{for } x \in \mathcal{D}(A), \tag{16.10}$$

where A_λ is the Yosida approximation of the operator A.

Proof. Application of Lemma 16.1 gives

$$\lim_{\lambda\to\infty} A_\lambda x = \lim_{\lambda\to\infty} \lambda R(\lambda, A)Ax = Ax \quad \text{for } x \in \mathcal{D}(A).$$ □

Lemma 16.3. *Let an operator $A: B \supset \mathcal{D}(A) \to B$ satisfy conditions (a) and (b) of Theorem* 16.1, *and $A_\lambda: B \to B$ be the Yosida approximation of the operator A. Then A_λ is the generator of the uniformly continuous contractive semigroup e^{tA_λ}, and for any $x \in B$ and $\lambda, \mu > 0$ the following estimate holds:*

$$\|e^{tA_\lambda}x - e^{tA_\mu}x\| \leqslant t\|A_\lambda x - A_\mu x\|. \tag{16.11}$$

Proof. From (16.9) it is clear that $A_\lambda: B \to B$ is a bounded linear operator. Consequently, by Theorem 14.2, the operator A_λ is the generator of a uniformly continuous semigroup e^{tA_λ} of bounded linear operators. Using (16.9) and (16.1), we arrive at the inequality

$$\|e^{tA_\lambda}\| = e^{-t\lambda}\|e^{t\lambda^2 R(\lambda,A)}\| \leqslant e^{-t\lambda}e^{t\lambda^2\|R(\lambda,A)\|} \leqslant 1. \tag{16.12}$$

Therefore, the semigroup e^{tA_λ} is contractive.

By definition, the operators e^{tA_λ}, e^{tA_μ}, A_λ and A_μ are mutually permutable. Taking this into account, we have

$$\|e^{tA_\lambda}x - e^{tA_\mu}x\| = \left\| \int_0^1 \frac{d}{ds}(e^{tsA_\lambda}e^{t(1-s)A_\mu}x)\, ds \right\| \leqslant$$
$$\leqslant \int_0^1 t\|e^{tsA_\lambda}e^{t(1-s)A_\mu}(A_\lambda x - A_\mu x)\|\, ds \leqslant t\|A_\lambda x - A_\mu x\|. \qquad \square$$

Proof of Theorem 16.1 (*sufficiency*). 1. Take an arbitrary $T_0 > 0$. It follows from (16.11) that

$$\|e^{tA_\lambda}y - e^{tA_\mu}y\| \leqslant t\|A_\lambda y - A_\mu y\| \leqslant t\|A_\lambda y - Ay\| + t\|Ay - A_\mu y\| \tag{16.13}$$

for all $y \in \mathcal{D}(A)$, $\lambda, \mu > 0$ and $t \in [0, T_0]$. Since $\mathcal{D}(A)$ is dense in B, for all $x \in B$ and $\varepsilon > 0$, there exists $x_\varepsilon \in \mathcal{D}(A)$ such that $\|x - x_\varepsilon\| < \varepsilon$. The inequality $\|e^{tA_\lambda}\| \leqslant 1$ gives

$$\|e^{tA_\lambda}x - e^{tA_\mu}x\| \leqslant \|e^{tA_\lambda}(x - x_\varepsilon)\| + \|e^{tA_\lambda}x_\varepsilon - e^{tA_\mu}x_\varepsilon\| +$$
$$+ \|e^{tA_\mu}(x - x_\varepsilon)\| < 2\varepsilon + \|e^{tA_\lambda}x_\varepsilon - e^{tA_\mu}x_\varepsilon\|.$$

Using (16.13) with $y = x_\varepsilon$ and Lemma 16.2, we obtain

$$\|e^{tA_\lambda}x_\varepsilon - e^{tA_\mu}x_\varepsilon\| \to 0 \quad \text{as } \lambda, \mu \to \infty$$

uniformly for $t \in [0, T_0]$, i. e., the uniform convergence of the family $\{e^{tA_\lambda}x\}$ on $[0, T_0]$. Let us denote the corresponding limit by $T(t)x$,

$$\lim_{\lambda\to\infty} e^{tA_\lambda}x = T(t)x \quad \text{for all } x \in B. \tag{16.14}$$

Convergence in (16.14) is uniform on any segment $[0, T_0]$.

It is clear that the mapping $T(t): B \to B$ is linear, $\|T(t)\| \leqslant 1$ and $T(0) = I$. Let us show that $T(t)$ satisfies the semigroup property. Based on this property for e^{tA_λ} and on the inequality $\|e^{tA_\lambda}\| \leqslant 1$, we have

$$\|T(t+s)x - T(t)T(s)x\| \leqslant \|T(t+s)x - e^{(t+s)A_\lambda}x\| +$$
$$+ \|e^{tA_\lambda}e^{sA_\lambda}x - e^{tA_\lambda}T(s)x\| + \|e^{tA_\lambda}T(s)x - T(t)T(s)x\| \leqslant$$
$$\leqslant \|T(t+s)x - e^{(t+s)A_\lambda}x\| + \|e^{sA_\lambda}x - T(s)x\| +$$
$$+ \|e^{tA_\lambda}(T(s)x) - T(t)(T(s)x)\| \to 0 \quad \text{as } \lambda \to \infty$$

for all $x \in B$, thus $T(t+s)x = T(t)T(s)x$.

Let us verify that $T(t)$ is a C_0-semigroup. Given $x \in B$ and $\varepsilon > 0$, we pick $\lambda > 0$ such that $\|T(t)x - e^{tA_\lambda}x\| < \varepsilon$ on the segment $t \in [0, T_0]$. Next, based on the λ found, choose $t_0 > 0$ such that $\|e^{tA_\lambda}x - x\| < \varepsilon$ for $0 < t < t_0$. Therefore,

$$\|T(t)x - x\| \leqslant \|T(t)x - e^{tA_\lambda}x\| + \|e^{tA_\lambda}x - x\| < 2\varepsilon \quad (0 < t < t_0).$$

This shows that $T(t): B \to B$ is a contractive C_0-semigroup.

2. It remains to prove that the operator $A: B \supset \mathcal{D}(A) \to B$ is the generator for $T(t)$. Let $x \in \mathcal{D}(A)$. Using the convergence of the family $\{e^{tA_\lambda}x\}$ to $T(t)x$ uniform on finite intervals, statement (d) of Theorem 15.3, and Lemma 16.2, we get

$$T(t)x - x = \lim_{\lambda\to\infty} (e^{tA_\lambda}x - x) = \lim_{\lambda\to\infty} \int_0^t e^{sA_\lambda} A_\lambda x\, ds = \int_0^t T(s)Ax\, ds. \tag{16.15}$$

Let $G: B \supset \mathcal{D}(G) \to B$ be the generator of the semigroup $T(t)$ and let $x \in \mathcal{D}(A)$. Dividing (16.15) by $t > 0$ and setting $t \downarrow 0$, we conclude based on statement (a) of Theorem 15.3 that $x \in \mathcal{D}(G)$ and $Gx = Ax$. Thus, $A \subset G$.

Since G is the generator of $T(t)$, it follows from condition (b) of Theorem 16.1 that $1 \in \rho(G)$. Therefore,

$$(I - G)\mathcal{D}(G) = B. \tag{16.16}$$

On the other hand, according the assumption in the theorem, $\mathbb{R}_+ \subset \rho(A)$, thus $1 \in \rho(A)$. Since $A \subset G$, we arrive at the equality

$$(I - G)\mathcal{D}(A) = (I - A)\mathcal{D}(A) = B. \tag{16.17}$$

It follows now from relations (16.16) and (16.17) that $\mathcal{D}(G) = (I - G)^{-1}B$ and $\mathcal{D}(A) = (I - G)^{-1}B$, i. e., $\mathcal{D}(A) = \mathcal{D}(G)$. We have $A = G$.

We have proven that every operator $A: B \supset \mathcal{D}(A) \to B$ satisfying conditions (a) and (b) of Theorem 16.1 is the generator of a contractive C_0-semigroup $T(t): B \to B, t \geqslant 0$. □

Remark 16.1. The first part of condition (a) in Theorem 16.1 is not an independent condition. In fact, the closedness of the operator A follows from condition (b). This conclusion is a consequence of the result given below.

Lemma 16.4. *Let $A: B \supset \mathcal{D}(A) \to B$ be a linear operator such that $\rho(A) \neq \varnothing$. Then A is closed.*

Proof. Let $w_k \in \mathcal{D}(A)$, $w_k \to w$ in B and $Aw_k \to y$ in B as $k \to \infty$. Let us show that $w \in \mathcal{D}(A)$ and $Aw = y$. Take $\lambda_0 \in \rho(A)$. Then $\lambda_0 w_k - Aw_k \to \lambda_0 w - y$ in B. In accordance with the definition of the resolvent $R(\lambda_0, A)$, we see that

$$w_k \to R(\lambda_0, A)(\lambda_0 w - y) \quad \text{as } k \to \infty.$$

The uniqueness of the limit implies

$$w = R(\lambda_0, A)(\lambda_0 w - y). \tag{16.18}$$

Therefore, $w \in \mathcal{D}(A)$. Applying the operator $(\lambda_0 I - A)$ to both sides of equation (16.18), we obtain

$$\lambda_0 w - Aw = \lambda_0 w - y,$$

i. e., $Aw = y$. The closedness of the operator A is proven. □

Corollary 16.1. *Let $A: B \supset \mathcal{D}(A) \to B$ be the generator of a contractive C_0-semigroup $T(t): B \to B$, $t \geqslant 0$ and $A_\lambda: B \to B$ be the Yosida approximation of the operator A. Then*

$$T(t)x = \lim_{\lambda\to\infty} e^{tA_\lambda}x \quad \textit{for } x \in B \textit{ and } t \geqslant 0. \tag{16.19}$$

Proof. The proof of Theorem 16.1 makes it clear that the expression on the right-hand side of (16.19) defines a contractive C_0-semigroup $S(t): B \to B$ with the generator A. Theorem 15.5 states that $T(t) = S(t)$, $t \geqslant 0$. □

Corollary 16.2. *Let $A: B \supset \mathcal{D}(A) \to B$ be the generator of a contractive C_0-semigroup $T(t): B \to B$, $t \geqslant 0$. Then $\{\lambda \in \mathbb{C}: \operatorname{Re}\lambda > 0\} \subset \rho(A)$ and*

$$\|R(\lambda, A)\| \leqslant \frac{1}{\operatorname{Re}\lambda} \quad \textit{for } \operatorname{Re}\lambda > 0. \tag{16.20}$$

Proof. If $\operatorname{Re}\lambda > 0$, then the integral on the right-hand side of (16.2) exists as an improper integral. Moreover, since $\|T(t)\| \leqslant 1$, we have

$$\|R(\lambda)x\| \leqslant \int_0^\infty e^{-\operatorname{Re}\lambda t}\|T(t)x\|\,dt \leqslant \frac{\|x\|}{\operatorname{Re}\lambda} \quad \text{for } \operatorname{Re}\lambda > 0.$$

Therefore, the operator $R(\lambda): B \to B$ is bounded and $\|R(\lambda)\| \leqslant 1/\operatorname{Re}\lambda$. Reasoning in the same way as in the proof of necessity in Theorem 16.1, we obtain $R(\lambda) = R(\lambda, A)$. □

Corollary 16.3. *A linear operator $A: B \supset \mathcal{D}(A) \to B$ is the generator of a C_0-semigroup $T(t): B \to B$ $(t \geqslant 0)$ satisfying the estimate*

$$\|T(t)\| \leqslant e^{\omega t} \quad (t \geqslant 0), \tag{16.21}$$

if and only if:
(a) *A is closed and $\overline{\mathcal{D}(A)} = B$.*
(b) *The resolvent set $\rho(A)$ contains the ray $\{\lambda \in \mathbb{R}: \lambda > \omega\}$ and*

$$\|R(\lambda, A)\| \leqslant \frac{1}{\lambda - \omega} \quad \textit{for all } \lambda > \omega. \tag{16.22}$$

Proof. 1. Let $A: B \supset \mathcal{D}(A) \to B$ be the generator of a C_0-semigroup $T(t): B \to B$ satisfying inequality (16.21). Consider the C_0-semigroup $S(t) = e^{-\omega t}T(t): B \to B$, $t \geqslant 0$. Inequality (16.21) gives

$$\|S(t)\| \leqslant e^{-\omega t}\|T(t)\| \leqslant e^{-\omega t}e^{\omega t} = 1.$$

It is easy to see that

$$\left.\frac{dS(t)x}{dt}\right|_{t=0} = (A - \omega I)x \quad (x \in \mathcal{D}(A)).$$

Consequently, the generator of the contractive semigroup $S(t)$ is the operator $G = A - \omega I$. Next, we obviously have

$$(\lambda I - A) = (\lambda - \omega)I - (A - \omega I) = (\lambda - \omega)I - G. \tag{16.23}$$

By the Hille–Yosida theorem, the subspace $\mathcal{D}(A) = \mathcal{D}(G)$ is dense in B and the operator $A = G + \omega I$ is closed. In addition, formula (16.23) means that $\lambda \in \rho(A)$ if and only if $\lambda - \omega \in \rho(G)$, i. e., we have $\{\lambda \in \mathbb{R}: \lambda > \omega\} \subset \rho(A)$ and

$$\|R(\lambda, A)\| \leqslant \frac{1}{\lambda - \omega} \quad \text{for } \lambda > \omega.$$

2. Now let an operator $A: B \supset \mathcal{D}(A) \to B$ satisfy conditions (a) and (b) of Corollary 16.3. Then the operator $G = A - \omega I$ satisfies conditions (a) and (b) of Theorem 16.1. Consequently, G is the generator of a contractive C_0-semigroup $S(t): B \to B$, $t \geqslant 0$. But then the operator $A = G + \omega I$ is the generator of the C_0-semigroup $T(t) = e^{\omega t}S(t)$ satisfying inequality (16.21). □

Example 16.1. Consider an operator $A: L_2(0,1) \supset \mathcal{D}(A) \to L_2(0,1)$ given by the formula

$$Ay = y'' + py' + qy,$$

with domain $\mathcal{D}(A) = \mathring{H}^1(0,1) \cap H^2(0,1)$, where $p, q \in \mathbb{R}$, $q \leqslant 0$.

Let us verify that A is the generator of a contractive C_0-semigroup.

1. Let us prove the inclusion $\mathbb{R}_+ \subset \rho(A)$. Consider the equation

$$\lambda y - Ay = f \tag{16.24}$$

and the corresponding homogeneous equation

$$\lambda w - Aw = 0, \tag{16.25}$$

where $\lambda > 0$ and $f \in L_2(0,1)$.

According to the definition of weak derivative, a function $y \in \mathcal{D}(A)$ is a solution to equation (16.24) if and only if y belongs to $\mathring{H}^1(0,1)$ and satisfies the integral identity

$$B(y, v) = \int_0^1 (y'\bar{v}' - py'\bar{v} + (\lambda - q)y\bar{v})\, dx = \int_0^1 f\bar{v}\, dx \tag{16.26}$$

for each function $v \in \mathring{H}^1(0,1)$. Similarly, equality (16.25) for $w \in \mathring{H}^1(0,1) \cap H^2(0,1)$ is equivalent to the integral identity

$$B(w,v) = \int_0^1 (w'\bar{v}' - pw'\bar{v} + (\lambda - q)w\bar{v})\,dx = 0 \tag{16.27}$$

with respect to $w \in \mathring{H}^1(0,1)$ for all $v \in \mathring{H}^1(0,1)$.

Let us prove that homogeneous equation (16.25) has the only trivial solution. In (16.27), take $v = w$. Since $\operatorname{Re}\int_0^1 w'\bar{w}\,dx = 0$, we have

$$\operatorname{Re} B(w,w) = \|w'\|^2_{L_2(0,1)} + (\lambda - q)\|w\|^2_{L_2(0,1)} \geqslant \lambda\|w\|^2_{L_2(0,1)}. \tag{16.28}$$

Relations (16.27) and (16.28) together mean that $w = 0$. Using Theorem 7.4, we make sure that equation (16.24) has a unique solution $y \in \mathring{H}^1(0,1) \cap H^2(0,1)$ for any function $f \in L_2(0,1)$. Therefore, the operator $\lambda I - A$ maps $\mathring{H}^1(0,1) \cap H^2(0,1)$ continuously and one-to-one onto the entire space $L_2(0,1)$. Therefore, by the Banach inverse operator theorem, there exists a bounded inverse operator $(\lambda I - A)^{-1}: L_2(0,1) \to \mathring{H}^1(0,1) \cap H^2(0.1)$.

2. Part 1 of the proof and Lemma 16.4 show that the operator $A: L_2(0,1) \supset \mathcal{D}(A) \to L_2(0,1)$ is closed.

3. By Theorem 4.4, the set $C_0^\infty(0,1)$ is dense in $L_2(0,1)$. Since $C_0^\infty(0,1) \subset \mathcal{D}(A)$, the domain of $\mathcal{D}(A)$ is also dense in $L_2(0,1)$.

4. It remains to establish estimate (16.1). Using the Cauchy–Schwarz inequality and integration by parts, similar to (16.28), we obtain

$$\begin{aligned}\|(\lambda I - A)y\|_{L_2(0,1)}\|y\|_{L_2(0,1)} &\geqslant \operatorname{Re}((\lambda I - A)y, y)_{L_2(0,1)} = \\ &= \operatorname{Re} B(y,y) \geqslant \lambda\|y\|^2_{L_2(0,1)} \quad \text{for } y \in \mathring{H}^1(0,1) \cap H^2(0,1),\end{aligned}$$

i. e.,

$$\|y\|_{L_2(0,1)} \leqslant \frac{1}{\lambda}\|(\lambda I - A)y\|_{L_2(0,1)}. \tag{16.29}$$

Let us denote $f = (\lambda I - A)y$. Then $y = R(\lambda, A)f$, and it follows from (16.29) that

$$\|R(\lambda, A)f\|_{L_2(0,1)} \leqslant \frac{1}{\lambda}\|f\|_{L_2(0,1)}$$

for any function $f \in L_2(0,1)$. This is exactly (16.1).

17 Cauchy problem for operator differential equations

Let $A: B \supset \mathcal{D}(A) \to B$ be a linear operator in a Banach space B.

Consider the operator differential equation

$$\frac{du(t)}{dt} = Au(t) + f(t) \quad (0 < t < T_0) \tag{17.1}$$

with the initial condition

$$u(0) = \varphi. \tag{17.2}$$

Here, $f \in C([0, T_0], B)$ and $\varphi \in B$. Let $C([0, T_0], B)$ denote the space of all continuous B-valued functions on $[0, T_0]$ with the norm

$$\|f\|_{C([0,T_0],B)} = \sup_{0 \leqslant t \leqslant T_0} \|f(t)\|,$$

and $C^1([0, T_0], B)$ be the space of all continuously differentiable functions on $[0, T_0]$ with values in B.

Definition 17.1. A function $u \in C([0, T_0], B)$ is called a classical solution to problem (17.1) and (17.2) if:
1. $u(t)$ is continuously differentiable on $(0, T_0)$,
2. $u(t) \in \mathcal{D}(A)$ for all $t \in (0, T_0)$,
3. $u(t)$ satisfies equation (17.1) for $t \in (0, T_0)$,
4. $u(t)$ satisfies initial condition (17.2).

Theorem 17.1. *Let a linear operator $A: B \supset \mathcal{D}(A) \to B$ be the generator of a C_0-semigroup $T(t): B \to B$, $t \geqslant 0$. If problem (17.1), (17.2) has a classical solution $u \in C([0, T_0], B)$, then this solution is represented as follows:*

$$u(t) = T(t)\varphi + \int_0^t T(t-s)f(s)\, ds. \tag{17.3}$$

Proof. Let $g(s) = T(t-s)u(s)$ $(0 \leqslant s \leqslant t \leqslant T_0)$. By Theorem 15.2, $g \in C([0, t], B)$. On the other hand, it follows from statement (c) of Theorem 15.3 that $g \in C^1((0, t), B)$ and

$$\frac{dg(s)}{ds} = -T(t-s)Au(s) + T(t-s)\frac{du(s)}{ds} = T(t-s)f(s) \qquad (0 < s \leqslant t < T_0). \tag{17.4}$$

By Theorem 15.2, $T(t-s)f(s) \in C([0, t], B)$. Therefore, integrating (17.4), we get

$$g(t) - g(0) = \int_0^t T(t-s)f(s)\, ds. \tag{17.5}$$

Obviously, $g(t) = u(t)$ and $g(0) = T(t)u(0) = T(t)\varphi$, and it follows then from (17.5) that

$$u(t) - T(t)\varphi = \int_0^t T(t-s)f(s)\, ds.$$

□

The following statement obviously follows from Theorem 17.1.

Corollary 17.1. *Let a linear operator $A\colon B \supset \mathcal{D}(A) \to B$ be the generator of a C_0-semigroup $T(t)\colon B \to B$. Then problem* (17.1), (17.2) *has at most one classical solution.*

Theorem 17.2. *Let a linear operator $A\colon B \supset \mathcal{D}(A) \to B$ be the generator of a C_0-semigroup $T(t)\colon B \to B$, $t \geqslant 0$. Assume also that $f \in C^1([0,T_0],B)$ and $\varphi \in \mathcal{D}(A)$. Then there exists a classical solution $u \in C^1([0,T_0],B)$ to problem* (17.1), (17.2).

Proof. Consider two auxiliary problems,

$$\frac{dw(t)}{dt} = Aw(t) \quad (0 < t < T_0), \tag{17.6}$$

$$w(0) = \varphi \tag{17.7}$$

and

$$\frac{dv(t)}{dt} = Av(t) + f(t) \quad (0 < t < T_0), \tag{17.8}$$

$$v(0) = 0. \tag{17.9}$$

Theorem 15.2 and statement (c) of Theorem 15.3 show that $w(t) = T(t)\varphi$ is a classical solution to problem (17.6), (17.7) and $w \in C^1([0,T_0],B)$.

Make sure that the function $v(t) = \int_0^t T(t-s)f(s)\,ds$ is a classical solution to problem (17.8), (17.9). In doing this, we first show that the function $v(t)$ has a continuous derivative for $0 < t < T_0$. By virtue of Theorems 15.2 and 15.3(a), we have

$$\begin{aligned} \frac{v(t+h)-v(t)}{h} &= \frac{1}{h}\left(\int_0^{t+h} T(t-s+h)f(s)\,ds - \int_0^t T(t-s)f(s)\,ds\right) = \\ &= \frac{1}{h}\int_0^t T(t-s)(f(s+h)-f(s))\,ds + \frac{1}{h}\int_0^h T(t-s+h)f(s)\,ds \to \\ &\to \int_0^t T(t-s)f'(s)\,ds + T(t)f(0) \quad \text{as } h \to 0 \quad (0 < t < T_0). \end{aligned} \tag{17.10}$$

Obviously, the function on the right-hand side of (17.10) belongs to $C([0,T_0],B)$. It remains to prove that $v(t) \in \mathcal{D}(A)$ and equation (17.8) holds on the interval $0 < t < T_0$. Based on the semigroup property, relation (17.10) and Theorems 15.2, 15.3(a), we get

$$\begin{aligned} \frac{T(h)-I}{h}v(t) &= \frac{1}{h}\int_0^t T(t-s+h)f(s)\,ds - \frac{1}{h}\int_0^t T(t-s)f(s)\,ds = \\ &= \frac{1}{h}\int_0^{t+h} T(t-s+h)f(s)\,ds - \frac{1}{h}\int_0^t T(t-s)f(s)\,ds - \end{aligned}$$

$$-\frac{1}{h}\int_t^{t+h} T(t-s+h)f(s)\,ds = \frac{v(t+h)-v(t)}{h} - \frac{1}{h}\int_t^{t+h} T(t-s+h)f(s)\,ds \to$$
$$\to v'(t) - f(t) \quad \text{as } h \downarrow 0 \quad (0 < t < T_0).$$

Therefore, $v(t) \in \mathcal{D}(A)$ and $Av(t) = v'(t) - f(t)$ for $0 < t < T_0$.

Obviously, the function $u(t) = v(t) + w(t)$ is a classical solution to problem (17.1), (17.2). It is also easy to see that $u \in C^1([0, T_0], B)$ under the conditions of the theorem. □

18 Mixed problems for parabolic equations

This section is devoted to the application of the semigroup theory to the study of mixed problems for parabolic equations.

Consider the parabolic equation

$$u_t(x,t) = \Delta u(x,t) - \sum_{i=1}^{n} a_i(x) u_{x_i}(x,t) - a_0(x)u(x,t) + \\ + f(x,t) \quad ((x,t) \in \Omega_{T_0}) \tag{18.1}$$

with the boundary condition

$$u(x,t) = 0 \quad ((x,t) \in \Gamma_{T_0}) \tag{18.2}$$

and the initial condition

$$u(x,t)|_{t=0} = \varphi(x) \quad (x \in Q). \tag{18.3}$$

Here, $Q \subset \mathbb{R}^n$ is a bounded domain, $\partial Q \in C^2$, $\Omega_{T_0} = Q \times (0, T_0)$, $\Gamma_{T_0} = \partial Q \times (0, T_0)$, the coefficients $a_i \in C^1(\overline{Q})$, $i = 1, \dots, n$ and $a_0 \in C(\overline{Q})$ are real-valued functions, $f \in L_2(\Omega_{T_0})$ and $\varphi \in L_2(Q)$ are given complex-valued functions.

If $a_0(x) \equiv 0$ $(x \in \bar{Q})$, then we deal with the *diffusion equation*. If $a_i(x) \equiv 0$, $i = 1, \dots, n$ and $a_0(x) \equiv 0$ $(x \in \bar{Q})$, then we have the heat equation.

Let us introduce an unbounded operator $A: L_2(Q) \supset \mathcal{D}(A) \to L_2(Q)$ according to the formula

$$Av = \Delta v(x) - \sum_{i=1}^{n} a_i(x) v_{x_i}(x) - a_0(x)v(x) \tag{18.4}$$

with domain

$$\mathcal{D}(A) = \mathring{H}^1(Q) \cap H^2(Q). \tag{18.5}$$

Note that the operator A is not self-adjoint. Therefore, in contrast to the heat equation, direct application of the Fourier method to study this problem is impossible. For this purpose, we use the semigroup method. Let us first prove that the operator A is the generator of a C_0-semigroup.

Theorem 18.1. *The operator $A: L_2(Q) \supset \mathcal{D}(A) \to L_2(Q)$ given by formulas* (18.4) *and* (18.5) *is the generator of a C_0-semigroup $T(t)$, $t \geqslant 0$, in the space $L_2(Q)$, and the estimate $\|T(t)\| \leqslant e^{\omega t}$, $t \geqslant 0$ holds with $\omega = \max_{x\in\overline{Q}}(-a_0(x) + \frac{1}{2}\sum_{i=1}^n a_{ix_i}(x))$.*

Proof. 1. Let us show that $\{\lambda \in \mathbb{R}: \lambda > \omega\} \subset \rho(A)$. Assuming $\lambda > \omega$, consider the Dirichlet problem

$$- \Delta v(x) + \sum_{i=1}^{n} a_i(x) v_{x_i}(x) + (a_0(x) + \lambda) v(x) = F(x) \quad (x \in Q), \tag{18.6}$$

$$v|_{\partial Q} = 0 \tag{18.7}$$

and the corresponding homogeneous problem

$$- \Delta w(x) + \sum_{i=1}^{n} a_i(x) w_{x_i}(x) + (a_0(x) + \lambda) w(x) = 0 \quad (x \in Q), \tag{18.8}$$

$$w|_{\partial Q} = 0 \tag{18.9}$$

(see Section 7).

A function $v \in \mathring{H}^1(Q)$ is called a weak solution to problem (18.6), (18.7) if for any function $\xi \in \mathring{H}^1(Q)$ the following integral identity holds:

$$B(v, \xi) = \int_Q F\bar{\xi}\, dx, \tag{18.10}$$

where

$$B(v, \xi) = \int_Q \nabla v \nabla \bar{\xi}\, dx + \int_Q \left(\sum_{i=1}^{n} a_i v_{x_i} + (a_0 + \lambda) v \right) \bar{\xi}\, dx. \tag{18.11}$$

A weak solution to homogeneous problem (18.8), (18.9) is defined similarly.

Let us prove that homogeneous problem (18.8), (18.9) has the only trivial solution. Set $v = \xi = w$ in (18.11). From the proof of Corollary 7.1, it is clear that

$$\operatorname{Re} B(w, w) \geqslant \int_Q \left(\lambda + a_0 - \frac{1}{2} \sum_{i=1}^{n} a_{ix_i} \right) |w|^2 dx \geqslant (\lambda - \omega) \|w\|^2_{L_2(Q)}. \tag{18.12}$$

In (18.10), we set $F(x) = 0$ and $v = \xi = w$. Then, based on (18.12), we obtain $0 \geqslant (\lambda - \omega)\|w\|^2_{L_2(Q)}$, whence $w = 0$. The Fredholm property of problem (18.6), (18.7) (see Theorem 7.4) allows us to conclude that for any function $F \in L_2(Q)$ this problem has a unique weak solution $v \in \mathring{H}^1(Q)$. Since $\partial Q \in C^2$, Theorems 7.8 and 7.9 on the smoothness of weak solutions of elliptic problems imply that $v \in H^2(Q)$ and equation (18.6) holds a. e. in Q. Consequently, if $\lambda > \omega$, then for any function $F \in L_2(Q)$ there exists a unique solution $v \in \mathring{H}^1(Q) \cap H^2(Q)$ of the operator equation

$$(\lambda I - A)v = F. \tag{18.13}$$

Therefore, $\{\lambda \in \mathbb{R}: \lambda > \omega\} \subset \rho(A)$.

2. From Part 1 of the proof and Lemma 16.4, it follows that the operator $A: L_2(Q) \supset \mathcal{D}(A) \to L_2(Q)$ is closed.

3. By Theorem 4.4, $C_0^\infty(Q)$ is dense in $L_2(Q)$. The inclusion of $C_0^\infty(Q) \subset \mathcal{D}(A)$ implies that $\mathcal{D}(A)$ is dense in $L_2(Q)$.

4. In view of Corollary 16.3, it remains to prove inequality (16.22). Applying the Cauchy–Schwarz inequality and integration by parts, similar to (18.12), we have

$$\begin{aligned} \|(\lambda I - A)w\|_{L_2(Q)} \|w\|_{L_2(Q)} &\geqslant \operatorname{Re}((\lambda I - A)w, w)_{L_2(Q)} = \\ &= \operatorname{Re} B(w, w) \geqslant (\lambda - \omega)\|w\|^2_{L_2(Q)} \end{aligned} \tag{18.14}$$

for $w \in \mathring{H}^1(Q) \cap H^2(Q)$.

Let us denote $f = (\lambda I - A)w$. Then $w = R(\lambda, A)f$, and (18.14) yields

$$\|R(\lambda, A)f\|_{L_2(Q)} \leqslant \frac{1}{\lambda - \omega} \|f\|_{L_2(Q)}$$

for any function $f \in L_2(Q)$, which is (16.22). □

From viewpoint of applications of the semigroup theory, it is natural to define a solution of parabolic problem (18.1)–(18.3) as the classical solution of the Cauchy problem for an operator differential equation in the sense of the previous section,

$$\frac{du}{dt} = Au + f(t), \tag{18.15}$$

$$u(0) = \varphi. \tag{18.16}$$

From Theorems 18.1 and 17.2, we immediately obtain the following statement.

Theorem 18.2. *Let $f \in C^1([0, T_0], L_2(Q))$ and $\varphi \in H^2(Q) \cap \mathring{H}^1(Q)$. Then problem* (18.15), (18.16) *has a unique classical solution u. This solution belongs to the space $C^1([0, T_0], L_2(Q))$ and is given by the formula*

$$u(t) = T(t)\varphi + \int_0^t T(t - s)f(s)\, ds. \tag{18.17}$$

Similar to the anisotropic space $H^{1,0}(\Omega_{T_0})$, we can introduce the space $H^{2,1}(\Omega_{T_0})$.

Definition 18.1. The space $H^{2,1}(\Omega_{T_0})$ is the linear space of complex-valued functions $u \in L_2(\Omega_{T_0})$ that have all weak derivatives $D_x^\alpha D_t^\beta u \in L_2(\Omega_{T_0})$ ($|\alpha| + 2\beta \leqslant 2$), with the inner product

$$(u,v)_{H^{2,1}(\Omega_{T_0})} = \sum_{|\alpha|+2\beta\leqslant 2} \int_{\Omega_{T_0}} D_x^\alpha D_t^\beta u\, D_x^\alpha D_t^\beta \bar{v}\, dxdt.$$

Corollary 18.1. *Under the conditions of Theorem* 18.2, *the function u in formula* (18.17) *belongs to the space* $H^{2,1}(\Omega_{T_0})$ *and is a weak solution to problem* (18.1)–(18.3).

Proof. So, the fact that $u \in H^{2,1}(\Omega_{T_0})$ means the existence of weak derivatives $u_t, u_{x_i}, u_{x_ix_j} \in L_2(\Omega_{T_0})$.

We have $Au = u' - f \in C([0,T_0], L_2(Q))$ for the classical solution u of problem (18.15), (18.16). Then $(\lambda I - A)u \in C([0,T_0], L_2(Q))$ for any fixed number $\lambda > \omega$. But, as was already seen in the proof of Theorem 18.1, the mapping $(\lambda I - A)^{-1}$ is bounded from $L_2(Q)$ to $H^2(Q)$. Therefore, $u \in C([0,T_0], H^2(Q))$. Consequently, the function u has all weak derivatives $u_{x_i}, u_{x_ix_j} \in L_2(\Omega_{T_0})$ (see Remark 11.1). The existence of a weak derivative $u_t = u'$ follows directly from the fact that u is continuously differentiable on $[0,T_0]$ as a function with values in $L_2(Q)$. In addition, the above mentioned weak derivatives are tied up by equation (18.1). Therefore, the integral identity

$$\int_{\Omega_{T_0}} \left(\nabla u \nabla \bar{v} - u\bar{v}_t + \left(\sum_{i=1}^n a_i u_{x_i} + a_0 u \right) \bar{v} \right) dxdt = \int_{\Omega_{T_0}} f\bar{v}\, dxdt + \int_Q \varphi \bar{v}|_{t=0} dx$$

for a weak solution can be obtained from (18.1) by multiplication by an arbitrary function $v \in H^1(\Omega_{T_0})$ such that $v|_{Q_{T_0} \cup \Gamma_{T_0}} = 0$, with the subsequent integration by parts.

Since u is a continuous function in t with values in $\mathring{H}^1(Q)$, it follows from the proofs of Theorems 6.7 and 11.3 that $u|_{\Gamma_{T_0}} = 0$. □

19 Cauchy problem for parabolic equations

In this section, we apply the semigroup theory to the Cauchy problem

$$u_t = \Delta u - \sum_{j=1}^n a_j u_{x_j} - a_0 u + f(x,t) \quad (x \in \mathbb{R}^n,\ t > 0), \tag{19.1}$$

$$u|_{t=0} = \varphi(x) \quad (x \in \mathbb{R}^n), \tag{19.2}$$

where it is assumed that $a_j \in \mathbb{R}$ ($j = 0, 1, \ldots, n$) and $a_0 \geqslant 0$.

Let us introduce an unbounded operator $A : L_2(\mathbb{R}^n) \supset \mathcal{D}(A) \to L_2(\mathbb{R}^n)$ according to the formula

$$Aw = \Delta w - \sum_{j=1}^{n} a_j w_{x_j} - a_0 w, \quad w \in \mathcal{D}(A) = H^2(\mathbb{R}^n). \tag{19.3}$$

Since $C_0^\infty(\mathbb{R}^n) \subset H^2(\mathbb{R}^n)$ and $C_0^\infty(\mathbb{R}^n)$ is dense in $L_2(\mathbb{R}^n)$, the operator A is densely defined.

Theorem 19.1. *The operator A defined by formulas* (19.3) *is the generator of a contractive C_0-semigroup $T(t) : L_2(\mathbb{R}^n) \to L_2(\mathbb{R}^n)$, $t \geqslant 0$.*

Proof. According to the Hille–Yosida theorem, it suffices to check that the equation

$$\lambda w - Aw = g \in L_2(\mathbb{R}^n) \tag{19.4}$$

has a unique solution $w \in H^2(\mathbb{R}^n)$ for positive λ and this solution satisfies the estimate

$$\|w\|_{L_2(\mathbb{R}^n)} \leqslant \frac{1}{\lambda}\|g\|_{L_2(\mathbb{R}^n)}. \tag{19.5}$$

In order to do this, let us move on to the Fourier image $\widehat{w}(\xi)$ of the function $w(x)$. Based on the Plancherel theorem and the formula for the Fourier image of the derivative, we see that the condition

$$\int_{\mathbb{R}^n} (1 + |\xi|^2)^2 |\widehat{w}(\xi)|^2 \, d\xi < \infty \tag{19.6}$$

is necessary and sufficient for the function w to belong to the space $H^2(\mathbb{R}^n)$, while the integral in (19.6) is the square of the equivalent norm in the space $H^2(\mathbb{R}^n)$. Moreover, if $w \in H^2(\mathbb{R}^n)$, then the Fourier image of the function w_{x_j} is the function $i\xi_j\widehat{w}$, and the Fourier image of Δw is $-|\xi|^2\widehat{w}$. Keeping this in mind, after the Fourier transform in equation (19.4) we obtain the following equation for $\widehat{w}(\xi)$:

$$\Big(\lambda + a_0 + i\sum a_j\xi_j + |\xi|^2\Big)\widehat{w}(\xi) = \widehat{g}(\xi), \tag{19.7}$$

and

$$\Big|\lambda + a_0 + |\xi|^2 + i\sum a_j\xi_j\Big|^2 = (\lambda + a_0 + |\xi|^2)^2 + \Big(\sum a_j\xi_j\Big)^2 \geqslant \lambda^2.$$

Moreover,

$$\Big|\lambda + a_0 + |\xi|^2 + i\sum a_j\xi_j\Big| \geqslant \lambda + a_0 + |\xi|^2 \geqslant c(1 + |\xi|^2),$$
$$c = \min(1, \lambda + a_0) > 0.$$

Consequently, the function $\widehat{w}$ determined uniquely by equation (19.7) satisfies the estimates

$$|\widehat{w}(\xi)| \leqslant |\widehat{g}(\xi)|/\lambda, \quad |\widehat{w}(\xi)| \leqslant \frac{|\widehat{g}(\xi)|}{c(1+|\xi|^2)} \quad (\xi \in \mathbb{R}^n).$$

The first estimate implies (19.5) and the second gives

$$\int_{\mathbb{R}^n} (1+|\xi|^2)^2 |\widehat{w}(\xi)|^2 \, d\xi \leqslant \frac{1}{c^2} \int_{\mathbb{R}^n} |\widehat{g}(\xi)|^2 \, d\xi,$$

i. e., $w \in H^2(\mathbb{R}^n)$ and the operator $\lambda I - A$ maps $H^2(\mathbb{R}^n)$ homeomorphically onto the entire space $L_2(\mathbb{R}^n)$. □

Let us now rewrite problem (19.1), (19.2) in the form of an operator differential equation

$$\frac{du}{dt} = Au + f(t) \tag{19.8}$$

with the initial condition

$$u(0) = \varphi. \tag{19.9}$$

From Theorems 19.1 and 17.2, we obtain the following result.

Theorem 19.2. *Suppose that $f \in C^1([0,T_0], L_2(\mathbb{R}^n))$ and $\varphi \in H^2(\mathbb{R}^n)$. Then problem* (19.8), (19.9) *has a unique classical solution u. This solution belongs to the space $C^1([0,T_0], L_2(\mathbb{R}^n))$ and is represented as follows:*

$$u(t) = T(t)\varphi + \int_0^t T(t-s)f(s)\, ds.$$

Remark 19.1. In accordance with the definition of a classical solution to problem (19.8) and (19.9), the values of $u(t)$ belong to $H^2(\mathbb{R}^n)$ for all $t \in (0, T_0)$. But in fact it is clear that $u \in C([0,T_0], H^2(\mathbb{R}^n))$. To make sure of this, note that if $u, f \in C^1([0,T_0], L_2(\mathbb{R}^n))$, then

$$(I-A)u = u - u' + f \in C([0,T_0], L_2(\mathbb{R}^n)).$$

Since $I - A$ is a linear homeomorphism of the spaces $H^2(\mathbb{R}^n)$ and $L_2(\mathbb{R}^n)$, we obtain the required statement.

Suppose now that $f = 0$ and $\varphi \in H^4(\mathbb{R}^n) = \mathcal{D}(A^2)$. We have

$$u'(t) = Au(t) = AT(t)\varphi = T(t)A\varphi,$$

where $A\varphi \in \mathcal{D}(A)$. In this case, there exists the second continuous derivative

$$u'' = ATA\varphi = TA^2\varphi \in C([0,+\infty), L_2(\mathbb{R}^n)),$$

i. e., $u \in C^2([0, +\infty), L_2(\mathbb{R}^n))$. It turns out that the smoothness of the initial function with respect to x ensures the smoothness of the solution with respect to t. Next, we deduce in a similar way from the conditions $u', u'' \in C([0, +\infty), L_2(\mathbb{R}^n))$ that $u, u' \in C([0, +\infty), H^2(\mathbb{R}^n))$, and then $u \in C([0, +\infty), H^4(\mathbb{R}^n))$.

As a good exercise, we recommend the reader to obtain an explicit representation for the semigroup $T(t)$ generated by the operator A in the case where $a_i = a_0 = 0$. It is clear that the result should be the Poisson integral, but the goal is to obtain a rigorous proof based on the semigroup theory.

Exercises

1) Construct an example of a bounded domain $Q \subset \mathbb{R}^n$ such that its boundary ∂Q has positive n-dimensional Lebesgue measure.

Hint. Use the procedure for constructing a set of Cantor type with positive measure.

2) For which values of $\delta \in \mathbb{R}$ does the function $u(x,y) = x^\delta$ belong to the space $L_p(Q)$ ($p \geqslant 1$), where $Q = \{(x,y) \in \mathbb{R}^2 : |y| < 1, |y|^\gamma < x < 1\}$, and $0 < \gamma \leqslant 1$?

3) A nonnegative function $f \in L_1(0,1)$ is such that $\int_a^b f\,dx > 0$ for all $0 \leqslant a < b \leqslant 1$. Is it true that $f > 0$ a. e. on $(0,1)$?

Hint. Construct a set of Cantor type $K \subset [0,1]$ with measure $\mu(K) = 1/2$. Set $f(x) = 1$ $(x \in [0,1] \setminus K)$, $f(x) = 0$ $(x \in K)$.

4) Prove that if a function $f \in L_{1,\text{loc}}(\mathbb{R}^n)$ satisfies the relation $\int_{\mathbb{R}^n} fg\,dx = 0$ for all $g \in C_0^\infty(\mathbb{R}^n)$, then $f = 0$ almost everywhere.

Hint. Take a mollifier as g and apply a result similar to Theorem 4.3, replacing the mean-square convergence with convergence in the mean.

5) Prove that the set $C_0^\infty(\mathbb{R}^n)$ is dense in $L_1(\mathbb{R}^n)$.

Hint. For an arbitrary function $f \in L_1(\mathbb{R}^n)$, consider the truncated function $f_R(x)$ coinciding with $f(x)$ for $|x| < R$ and equal to zero for $|x| \geqslant R$, then use the averaging procedure.

6) Consider the function $u(x_1,x_2) = \operatorname{sign} x_1 + \operatorname{sign} x_2$ in $Q = \{x \in \mathbb{R}^2 : x_1^2 + x_2^2 < 1\}$. Prove the existence of the weak derivative $u_{x_1x_2}$ in Q.

Hint. Verify that the weak derivative $(\operatorname{sign} x_i)_{x_j}$, $i \neq j$, exists and is equal to zero.

7) Prove that the trace of a function defined on page 59 does not depend on the choice of an approximating sequence.

Hint. Consider two different sequences $u_m, \tilde{u}_m \in C^1(\overline{Q})$, converging to u in $H^1(Q)$ and generating traces $u|_{\partial Q}$ and $\tilde{u}|_{\partial Q}$, then apply the triangle inequality to estimate $\|u|_{\partial Q} - \tilde{u}|_{\partial Q}\|_{L_2(\partial Q)}$.

8) Prove that the trace space $H^{1/2}(\partial Q)$ with norm (6.7) is a Banach space.

Hint. Use the definition of infimum norm (6.7) and the completeness of the space $H^1(Q)$.

9) Let a family of functions $\mathcal{F} \subset C_0^1(\mathbb{R}^n)$ be such that (1) supports of all functions in the family are contained in a fixed ball; (2) the set of integrals

$$\int_{\mathbb{R}^n} (1 + |\xi|^2)^{1/2} |\hat{v}(\xi)|^2\,d\xi \quad (v \in \mathcal{F})$$

is bounded. Prove that $\mathcal{F}$ is precompact in $L_2(\mathbb{R}^n)$.

https://doi.org/10.1515/9783112229637-006

Hint. Apply reasoning similar to that one used in the proof of Theorem 6.5.

10) For functions $f \in C_0^1(\mathbb{R}^n)$, $f = f(x', x_n)$, prove the estimate

$$\int\limits_{\mathbb{R}^{n-1}} (1 + |\xi'|^2)^{1/2} |\hat{f}'(\xi', 0)|^2 \, d\xi' \leqslant C \|f\|^2_{H^1(\mathbb{R}^n)},$$

where the constant $C > 0$ does not depend on f, and

$$\hat{f}'(\xi', 0) = (2\pi)^{-(n-1)/2} \int\limits_{\mathbb{R}^{n-1}} f(x', 0) e^{-i(\xi', x')} \, dx'$$

is the $(n-1)$-dimensional Fourier transform of the restriction of the function f to the hyperplane $x_n = 0$.

Hint. Use the representation $\hat{f}'(\xi', 0) = (2\pi)^{-1/2} \int_{-\infty}^{+\infty} \hat{f}(\xi', \xi_n) \, d\xi_n$ and the Plancherel theorem.

11) Prove that the trace operator from $H^1(Q)$ to $L_2(\partial Q)$ is compact.

Hint. Based on a suitable partition of unity in $\overline{Q}$ and local straightening of ∂Q, reduce the problem to Exercises 9 and 10.

12) Let B_1 be the open unit disk on the plane and

$$a_n = \frac{1}{\sqrt{2\pi}} \int\limits_0^{2\pi} \gamma(\varphi) e^{-in\varphi} \, d\varphi \quad (n \in \mathbb{Z})$$

be the Fourier coefficients of an $L_2(\partial B_1)$-function γ in the orthonormal basis $\{(2\pi)^{-1/2} e^{in\varphi}\}_{n \in \mathbb{Z}}$ in $L_2(\partial B_1)$. Prove that the condition

$$\sum_{n=-\infty}^{+\infty} n|a_n|^2 < \infty \qquad (19.10)$$

is necessary and sufficient for the function γ to belong to the trace space $H^{1/2}(\partial B_1)$.

Prove that the operator reconstructing the harmonic function u in B_1 from its trace γ on ∂B_1 is bounded from $H^{1/2}(\partial B_1)$ to $H^1(B_1)$.

Hint. Consider the function $v = \sum_{n=-\infty}^{+\infty} a_n r^{|n|} e^{in\varphi}$ harmonic in B_1, belonging to the space $H^1(B_1)$ if (19.10) holds, and such that $v|_{\partial B_1} = \gamma$. In order to prove the necessity of inequality (19.10), consider the Bessel inequality for an arbitrary function from $H^1(B_1)$ with respect to the orthogonal system $\{r^{|n|} e^{in\varphi}\}_{n \in \mathbb{Z}}$ in $H^1(B_1)$ with the inner product

$$(u, v)_{H^1(B_1)} = \iint\limits_{B_1} \nabla u \nabla \overline{v} \, dx_1 dx_2 + \int\limits_0^{2\pi} u|_{\partial B_1} \overline{v}|_{\partial B_1} \, d\varphi.$$

13) Prove that the space $H^1(Q)$ is not contained in $C(Q)$ for a bounded domain $Q \subset \mathbb{R}^2$.

Hint. Show that the function $\ln|\ln|x||$ belongs to $H^1(B_{1/2})$.

14) Prove that the space $H^1(a,b)$ coincides with the space of all absolutely continuous functions on $[a,b]$ whose derivatives belong to $L_2(a,b)$.

Hint. Make sure that any function f from $H^1(a,b)$ is representable as a Lebesgue integral with a variable upper limit, $f(x) = f(x_0) + \int_{x_0}^{x} g(s)\,ds$, where $g \in L_2(a,b)$, and x_0 is an arbitrary point from $[a,b]$.

15) For which values of $\alpha \in \mathbb{R}$ and $\beta \in \mathbb{R}$ does the function $u(r,\varphi) = r^\alpha \sin\beta\varphi$ belong to the space (a) $H^1(B_1)$; (b) $H^2(B_1)$? Here, r and φ are the polar coordinates on the plane, and B_1 is the open unit disk $\{r < 1\}$. Under which conditions on α and β the function $u \in H^1(B_1)$ is a weak solution to the problem

$$\Delta u = 0 \quad (0 < r < 1,\ 0 < \varphi < \omega),$$
$$u|_{\varphi=0} = u|_{\varphi=\omega} = 0?$$

Hint. Express the norms in the spaces $H^1(B_1)$ and $H^2(B_1)$, as well as the Laplace operator, in the polar coordinates, use the reasoning from Examples (5.1), (5.3).

16) For which values of $\alpha \in \mathbb{R}$ does the function $u(x,y) = |\ln(x^2 + xy + 2y^2)|^\alpha$ belong to the space $H^1(\Omega)$, where $\Omega = (-1/4, 1/4) \times (-1/4, 1/4)$?

17) Specify the conditions on $a_k \in \mathbb{C}$ that are necessary and sufficient for the function

$$f(x,y) = \sum_{k=1}^{\infty} a_k e^{-ky} \sin kx$$

to belong to the space $H^1(\Pi)$, where Π is the half-strip $\{0 < x < \pi,\ y > 0\}$.

Hint. Use orthogonality of the series in $H^1(\Pi)$ to verify its convergence.

18) Let a domain Q be divided by a smooth surface S into two parts Q_1 and Q_2, and let a function $u(x)$ defined in Q be such that $u|_{Q_i} \in C^1(\overline{Q}_i)$, $i = 1,2$. Make sure that the continuity of u on S is sufficient for the existence of the first-order weak derivatives from $L_2(Q)$ of the function u in the domain Q. In which case is this continuity condition also necessary?

Hint. See the reasoning in Example (5.1).

19) Let $u \in H^k(Q_s)$, $Q_s = (s-1,s) \times (0,1)$, $s = 1,2$. Establish necessary and sufficient conditions for u to belong to $H^k(Q)$, where $Q = (0,2) \times (0,1)$.

20) Denote $D = \{x \in \mathbb{R}^n : x_1^2 + \cdots + x_{n-1}^2 < x_n^2,\ 0 < x_n < +\infty\}$. Prove the following statement: for all $C > 0$, there exists a bounded domain $\Omega \subset D$ and a function $f \in \mathring{H}^1(\Omega)$ such that

$$\int_\Omega |f|^2\,dx > C \int_\Omega |\nabla f|^2\,dx.$$

Hint. Take advantage of the fact that a cone contains a ball of an arbitrarily large radius.

21) Is the Friedrichs inequality valid in the strip

$$\Pi = \{(x, y) \in \mathbb{R}^2 : 0 < x < 1, -\infty < y < +\infty\}?$$

Hint. Rely on the Friedrichs inequality in $\mathring{H}^1(0,1)$.

22) Let $Q \subset \mathbb{R}^n$ be a bounded domain. Prove that each of the formulas

$$\text{(a)}\quad (u, v)'_{H^1(Q)} = \int_Q \nabla u \nabla \overline{v}\, dx + \int_Q u\, dx \cdot \int_Q \overline{v}\, dx,$$

$$\text{(b)}\quad (u, v)'_{H^1(Q)} = \int_Q \nabla u \nabla \overline{v}\, dx + \int_{\partial Q} u|_{\partial Q}\, dS \cdot \int_{\partial Q} \overline{v}|_{\partial Q}\, dS$$

defines an inner product in the space $H^1(Q)$ equivalent to the standard one. This implies, in particular, that the integral $\int_Q \nabla u \nabla \overline{v}\, dx$ determines the equivalent inner product in the closed subspaces

$$\left\{u \in H^1(Q) : \int_Q u\, dx = 0\right\}, \quad \left\{u \in H^1(Q) : \int_{\partial Q} u|_{\partial Q}\, dS = 0\right\}$$

of the space $H^1(Q)$.

Hint. Assume the opposite, namely that there exists a sequence $u_n \in H^1(Q)$ orthonormal in $L_2(Q)$ such that

$$\||\nabla u_n|\|^2_{L_2(Q)} + \left|\int_{\partial Q} u_n|_{\partial Q}\, ds\right|^2 < \frac{1}{n}$$

(for case (b)). Based on the compactness of the embedding of $H^1(Q)$ in $L_2(Q)$, arrive at a contradiction.

23) Let $\mathring{H}^1_S(Q)$ denote the (closed) subspace of $H^1(Q)$, consisting of all functions from $H^1(Q)$ with zero traces on a smooth part $S \subset \partial Q$ of the boundary. Prove that the formula

$$(u, v)_{\mathring{H}^1_S(Q)} = \int_Q \nabla u \nabla \overline{v}\, dx$$

defines an inner product in the space $\mathring{H}^1_S(Q)$ equivalent to the standard one induced from $H^1(Q)$.

24) For all $u \in H^k(Q)$ and $q > 0$, prove the *interpolation inequality*

$$q^{k-s}\|u\|_{H^s(Q)} \leqslant c_1\left(\|u\|_{H^k(Q)} + q^k\|u\|_{L_2(Q)}\right),$$

where $k, s \in \mathbb{N}$, $s < k$; $c_1 > 0$ is a constant independent of u and q.

Hint. Prove the interpolation inequality for functions defined in $\mathbb{R}^n$, rely on the Fourier transform. In the case of a bounded domain Q use the extension theorem.

25) For all $u \in H^1(Q)$ and $q > 0$, prove the *interpolation inequality*

$$q^{1/2}\|u|_{\partial Q}\|_{L_2(\partial Q)} \leqslant c_2(\|u\|_{H^1(Q)} + q\|u\|_{L_2(Q)}),$$

where $c_2 > 0$ is a constant independent of u and q.

Hint. First, prove the interpolation inequality in the case $Q = \mathbb{R}^n_+ = \{x = (x', x_n) \in \mathbb{R}^n : x_n > 0\}$, using the extension operator to $\mathbb{R}^n$ (bounded in L_2 and H^1) and the Fourier transform. The case of a bounded domain Q is reduced to the half-space $\mathbb{R}^n_+$ by means of a suitable partition of unity in $\overline{Q}$ and a local straightening of ∂Q.

26) Solve the following eigenfunction and eigenvalue problems:

(a) $-u''(x) = \lambda u(x) \quad (0 < x < \pi), \quad u(0) = u(\pi) = 0,$

(b) $-u''(x) = \lambda u(x) \quad (0 < x < \pi), \quad u'(0) = u'(\pi) = 0,$

(c) $-u''(x) = \lambda u(x) \quad (0 < x < \pi), \quad u(0) = u'(\pi) = 0,$

(d) $-u''(x) = \lambda u(x) \quad (0 < x < 2),$
$u(0) = \alpha u(1), \quad u(2) = \beta u(1)$
(inspect the dependence on $\alpha \in \mathbb{R}$ and $\beta \in \mathbb{R}$),

(e) $-\Delta u(x,y) = \lambda u(x,y) \quad (0 < x < a, 0 < y < b),$
$u|_{x=0} = u|_{x=a} = 0, \quad u|_{y=0} = u|_{y=b} = 0.$

(f) $-\Delta u(x,y) = \lambda u(x,y) \quad (0 < x < a, 0 < y < b),$
$u_x|_{x=0} = u_x|_{x=a} = 0, \quad u_y|_{y=0} = u_y|_{y=b} = 0.$

27) Let B_1 be the unit disk on the plane, Γ be the circle of radius 1/2 centered at the origin, and f be a distribution in B_1 given by the formula

$$f(v) = \int_\Gamma v\,dS \quad (v \in C_0^\infty(B_1)).$$

Show that the Dirichlet problem

$$-\Delta u = f \quad \text{in } B_1, \quad u|_{\partial B_1} = 0$$

has a unique weak solution $u \in \mathring{H}^1(B_1)$. Find this solution.

Hint. Make sure that f extends to a continuous linear functional on $\mathring{H}^1(B_1)$. Obtain the integral identity for a weak solution of the problem, use the reasoning of Theorem 7.1. In order to find this solution $u \in \mathring{H}^1(B_1)$, note that the restrictions of u to the disk $\{|x| < 1/2\}$ and the ring $\{1/2 < |x| < 1\}$ are harmonic functions depending on $|x|$.

28) Make sure that Theorem 7.1 remains valid even without the assumption of smoothness of the boundary (in this case, of course, the connection between the weak solution and the classical formulation of the Dirichlet problem becomes more formal, in particular, note the loss of the description of the space $\mathring{H}^1(Q)$ as the set of functions from $H^1(Q)$ whose traces on ∂Q are equal to zero).

Hint. Show that the space $\mathring{H}^1(Q)$ is compactly embedded in the space $L_2(Q)$ for any bounded domain $Q \subset \mathbb{R}^n$ whose boundary has zero n-dimensional Lebesgue measure.

29) Let $Q \subset \mathbb{R}^n$ be a bounded domain, $\partial Q \in C^1$, and $\widetilde{L}_2(\partial Q)$ the subspace $\{\psi \in L_2(\partial Q) : \int_{\partial Q} \psi \, dS = 0\}$. Verify that for any function $\psi \in \widetilde{L}_2(\partial Q)$ the Neumann problem

$$-\Delta u = 0 \quad \text{in } Q, \quad \left.\frac{\partial u}{\partial \nu}\right|_{\partial Q} = \psi$$

has a unique weak solution $u \in H^1(Q)$ such that $\varphi = u|_{\partial Q} \in \widetilde{L}_2(\partial Q)$. Therefore, a bounded linear operator $A : \widetilde{L}_2(\partial Q) \to \widetilde{L}_2(\partial Q)$, $A\psi = \varphi$, arises. Show that the spectrum of A consists of isolated eigenvalues $\lambda_s > 0$ of finite multiplicity, $\lambda_s \to 0$ as $s \to \infty$, and the corresponding eigenfunctions ψ_s form an orthonormal basis of the space $\widetilde{L}_2(\partial Q)$.

Hint. Make sure that A is a positive compact operator in $\widetilde{L}_2(\partial Q)$, and apply the Hilbert–Schmidt theorem.

30) Solve the boundary-value problem

$$\begin{aligned} -\Delta u &= \sin x_1 \cos 4x_2 \quad (0 < x_1 < \pi, 0 < x_2 < \pi), \\ u|_{x_1=0} &= u|_{x_1=\pi} = 0, \\ u_{x_2}|_{x_2=0} &= u_{x_2}|_{x_2=\pi} = 0. \end{aligned}$$

Hint. Use the separation of variables x_1 and x_2 analogous to the Fourier method outlined for mixed problems in Chapter 4.

31) Solve the boundary-value problem

$$\begin{aligned} -\Delta u &= r \quad (1 < r < 2,\ 0 < \varphi < \pi), \\ u|_{r=1} &= \cos 2\varphi, \quad u|_{r=2} = 0, \\ \left.\frac{\partial u}{\partial \nu}\right|_{\varphi=0} &= \left.\frac{\partial u}{\partial \nu}\right|_{\varphi=\pi} = 0. \end{aligned}$$

Hint. Use the separation of variables in the polar coordinates. Search for a solution in the form of a series of eigenfunctions of the boundary-value problem $-e''(\varphi) = \lambda e(\varphi)$ $(0 < \varphi < \pi)$, $e'(0) = e'(\pi) = 0$.

32) Let (r, φ) be the polar coordinates on the plane. For which values of α and $\beta \in \mathbb{R}$ is the Neumann problem

$$-\Delta u = r^2 \sin^2 \varphi \quad (1 < r < 2),$$
$$\left.\frac{\partial u}{\partial \nu}\right|_{r=1} = \alpha \sin^2 \varphi, \quad \left.\frac{\partial u}{\partial \nu}\right|_{r=2} = \beta \cos^2 \varphi$$

solvable?

Hint. Use Theorem 8.1.

33) Consider the problem

$$-\sum_{i,j=1}^{n} (R_{ij} u_{x_j})_{x_i} = f(x) \quad (x \in Q), \tag{1}$$

$$u|_{\partial Q} = 0, \tag{2}$$

where $R_{ij} : L_2(Q) \to L_2(Q)$ are bounded linear operators such that the Hermitian part of the matrix operator

$$\mathbf{R} = (R_{ij})_{i,j=1}^{n} : L_2^n(Q) \to L_2^n(Q)$$

is positive-definite:

$$\mathrm{Re}\,(\mathbf{R}V, V)_{L_2^n(Q)} \geqslant c_0 \|V\|^2_{L_2^n(Q)} \quad (V \in L_2^n(Q)),$$

where $L_2^n(Q) = L_2(Q) \times \cdots \times L_2(Q)$.

Formulate the definition of a weak solution to problems (1), (2). Prove that a weak solution exists and is unique for any function $f \in L_2(Q)$. Show that eigenvalues λ_s in this problem are isolated, have finite multiplicity, and $\mathrm{Re}\,\lambda_s > 0$.

34) Let (r, φ) be the polar coordinates on the plane and B_1 be the open unit disk $\{r < 1\}$. Inspect the solvability of the Poisson equation $-\Delta u = f$ in B_1 with the nonlocal boundary condition $u(1, \varphi) = bu(1/2, \varphi)$ depending on the parameter $b \in \mathbb{R}$, where $f \in L_2(B_1)$.

Hint. Search for a solution in the form of a series of eigenfunctions of the boundary-value problem $-e''(\varphi) = \lambda e(\varphi)$, $e(\varphi) \equiv e(\varphi + 2\pi)$.

35) Find

$$\inf_M \left\{ \int_\Omega (|\nabla u|^2 + 2u)\, dx + \int_{|x|=1} (u|_{|x|=1})^2\, dS \right\},$$

where $\Omega = \{1 < |x| < 2\}$, $x = (x_1, x_2, x_3)$ and $M = \{v \in H^1(\Omega) : v|_{|x|=2} = 0\}$.

Hint. Introduce the inner product in the space M by the formula

$$(u, v)_M = \int_\Omega \nabla u \nabla v\, dx + \int_{|x|=1} uv\, dS.$$

Applying the reasoning of Section 7 (p. 77), proceed to finding a weak solution to the boundary-value problem for the Poisson equation in the ring.

36) Find the minimum of the quadratic functional

$$J(u) = \iint_G (u_x^2(x,y) + u_y^2(x,y) + u_x(x-1,y)u_x(x,y) + u_y(x-1,y)u_y(x,y) + \\ + u(x,y)f(x,y))\,dxdy$$

for $u \in \mathring{H}^1(G)$, where $G = (0,2) \times (0,1)$ and $f \in L_2(G)$. It is assumed that $u(x,y) = 0$ if $(x,y) \in \mathbb{R}^2 \setminus G$.

Hint. 1) Make sure that the quadratic part of $J(u)$ is the square of an equivalent norm in the space $\mathring{H}^1(G)$.

2) Reasoning similarly to Section 7 (p. 77), proceed to the integral identity

$$\iint_G ((Ru)_x v_x + (Ru)_y v_y)\,dxdy = -\frac{1}{2}\iint_G fv\,dxdy \quad (v \in \mathring{H}^1(G))$$

equivalent to the variational problem and defining a weak solution to the Dirichlet problem for the differential-difference equation

$$\Delta Ru(x,y) = f(x,y) \quad ((x,y) \in G),$$

where

$$Ru(x,y) = u(x,y) + \frac{1}{2}(u(x-1,y) + u(x+1,y)).$$

3) Show that a function w belongs to the image of the space $\mathring{H}^1(G)$ under the action of the difference operator R if and only if $w \in H^1(G)$ and

$$w|_{y=0} = w|_{y=1} = 0, \quad w|_{x=0} = \frac{1}{2}w|_{x=1} = w|_{x=2}. \tag{19.11}$$

4) Setting $w = Ru$, solve the (nonlocal) boundary-value problem for the Poisson equation $\Delta w = f$ in the domain G with boundary conditions (19.11).

37) Verify the validity of estimate (7.43) directly.

38) Give rigorous proofs of the statements in Remarks 7.3 and 7.5 concerning the smoothness of weak solutions of the Dirichlet problem.

39) Formulate and prove generalizations of Remarks 7.3 and 7.5 to the case of the Neumann and Robin problems.

40) Give an example of a function $f \in C(\overline{Q})$ such that problem (9.5), (9.6) does not have a classical solution (namely, consider a function f continuous in Q such that the equation $-\Delta u = f$ does not have twice continuously-differentiable solutions in Q).

Hint. Consider the function $u(x_1, x_2) = (x_1^2 - x_2^2)(-\ln|x|)^{1/2}$ in the disk $\{x \in \mathbb{R}^2 : x_1^2 + x_2^2 < R^2 < 1\}$.

41) Verify the existence of a classical solution to problem (9.5), (9.6) in the ball $Q = \{x \in \mathbb{R}^n : |x| < 1\}$ in the case where the function $f \in C(\overline{Q})$ depends only on $|x|$.

42) Prove Theorems 11.1 and 11.2.

43) Solve the following mixed problems for hyperbolic equations:

$$\begin{aligned} \text{a)}\quad & u_{tt} = u_{xx} + (4t+3)\sin 9x \quad (0 < x < \pi/6,\ t > 0), \\ & u|_{t=0} = 2, \quad u_t|_{t=0} = 0, \quad u|_{x=0} = u_x|_{x=\pi/6} = 0; \\ \text{b)}\quad & u_{tt} + 2u_t = u_{xx} - u \quad (0 < x < \pi,\ t > 0), \\ & u|_{x=0} = u|_{x=\pi} = 0, \quad u|_{t=0} = \pi x - x^2, \quad u_t|_{t=0} = 0. \end{aligned}$$

44) Prove that the set $\{u \in C^2(\overline{\Omega}_T) : u(x,t) = 0 \text{ on } \Gamma_T \cup Q_T\}$ is dense in $\widetilde{H}^1(\Omega_T)$.
Hint. See the proof of Theorem 6.7.

45) Prove Theorem 10.3 in the case $0 \neq \varphi \in \mathring{H}^1(Q)$ and $0 \neq \psi \in L_2(Q)$ (nonhomogeneous initial conditions).

Hint. Consider separately the nonhomogeneous wave equation with the homogeneous initial conditions and the homogeneous wave equation with the nonhomogeneous initial conditions. Following the calculations used in the proof of Theorem 10.3, estimate the resulting integral $\int_{Q_0}(w_{mt}^2 + |\nabla w_m|^2)\,dx$ as in Theorem 11.8.

46) Prove statement (b) on page 113, which is used to verify the validity of Remark 11.1.

Hint. Make sure that linear combinations of functions of the form $v_k(x)\theta(t)$ are dense in the space $\mathring{H}^1(\Omega_T)$, where $\theta \in C^1[0,T]$, $\theta(0) = \theta(T) = 0$, and the sequence $v_1, v_2, \ldots$ forms an orthonormal basis of the space $\mathring{H}^1(Q)$; see reasoning on page 110.

47) Prove Theorem 11.1.

48) Prove Theorem 11.2.

49) Prove Theorem 11.3.

50) Solve the mixed problem for the wave equation in a rectangle:

$$\begin{aligned} & u_{tt}(x,y,t) = \Delta u(x,y,t) \quad (0 < x < p,\ 0 < y < q,\ t > 0), \\ & u|_{x=0} = u|_{x=p} = 0, \quad u|_{y=0} = u|_{y=q} = 0, \\ & u|_{t=0} = Axy(x-p)(y-q), \quad u_t|_{t=0} = 0, \end{aligned}$$

where $p > 0$, $q > 0$, and A are constants.

51) Solve the following mixed problem for the heat equation:

$$u_t = u_{xx} \quad (0 < x < l,\ t > 0),$$

$$u_x|_{x=0} = u_x|_{x=l} = 0, \quad u|_{t=0} = \begin{cases} u_0, & 0 < x < l/2, \\ 0, & l/2 < x < l, \end{cases}$$

where $u_0 \neq 0$ is a constant.

52) Solve the following mixed problem for a parabolic equation:

$$u_t = u_{xx} - 8u_x + 3t^2(\pi - x) - 8t^3 \quad (0 < x < \pi,\ t > 0),$$
$$u|_{t=0} = 3e^{4x} \sin 2x, \quad u|_{x=0} = \pi t^3, \quad u|_{x=\pi} = 0.$$

Hint. Make a suitable change of the unknown function in order to free the equation from the first-order derivative with respect to x and make the boundary conditions homogeneous.

53) Solve the mixed problem for the heat equation in a ball:

$$u_t = \Delta u \quad (x \in \mathbb{R}^3 : r = |x| < R,\ t > 0),$$
$$(u_r + hu)|_{r=R} = hu_1, \quad u|_{t=0} = u_0,$$

where $h > 0$, $R > 0$, u_0, and u_1 are constants.

Hint. The solution is a function of $r = |x|$ and t.

54) Derive the D'Alembert formula.

55) Solve the following problems for the wave equation on the semiaxis:

a) $u_{tt} = u_{xx} + \sin x \quad (x > 0,\ t > 0),$
$u|_{t=0} = 0, \quad u_t|_{t=0} = xe^{-x} \quad (x > 0), \quad u|_{x=0} = 0 \quad (t > 0);$

b) $u_{tt} = u_{xx} + e^{-x} \quad (x > 0,\ t > 0),$
$u|_{t=0} = \sin x, \quad u_t|_{t=0} = 0 \quad (x > 0), \quad u_x|_{x=0} = 0 \quad (t > 0);$

c) $u_{tt} = u_{xx} \quad (x > 0,\ t > 0),$
$u|_{t=0} = \cos x, \quad u_t|_{t=0} = e^{-x} \quad (x > 0), \quad u|_{x=0} = 0 \quad (t > 0).$

Hint. Use a suitable extension of the data to the entire axis to apply the D'Alembert formula.

56) Solve the problem for the wave equation on the semiaxis ($a > 0$ is a constant)

$$u_{tt} = a^2 u_{xx} \quad (x > 0,\ t > 0),$$
$$u|_{t=0} = \varphi(x), \quad u_t|_{t=0} = \psi(x) \quad (x > 0), \quad u|_{x=0} = \alpha(t) \quad (t > 0),$$

where $\alpha(0) = \varphi(0)$.

57) Solve the problem for the wave equation on the semiaxis:

$$
\begin{aligned}
u_{tt} &= 4u_{xx} + 2 \quad (x > 0,\ t > 0),\\
u|_{t=0} &= 2x - \cos x, \quad u_t|_{t=0} = -2 - 2\sin x \quad (x > 0),\\
(u - u_x)|_{x=0} &= t^2 - 3 \quad (t > 0).
\end{aligned}
$$

58) Let $\psi(x) = 0$ for all $x \in \mathbb{R}^2$ such that $|x_1| + |x_2| > 1$. For all $t > 0$, find $x \in \mathbb{R}^2$ such that the solution to the Cauchy problem

$$
\begin{aligned}
u_{tt} &= \Delta u \quad (x \in \mathbb{R}^2,\ t > 0),\\
u|_{t=0} &= 0, \quad u_t|_{t=0} = \psi(x) \quad (x \in \mathbb{R}^2)
\end{aligned}
$$

is equal to zero.

59) Using the Kirchhoff formula, solve the Cauchy problem for the wave equation

$$
\begin{aligned}
u_{tt} &= \Delta u \quad (x \in \mathbb{R}^3,\ t > 0),\\
u|_{t=0} &= e^{-|x|^2}, \quad u_t|_{t=0} = 0 \quad (x \in \mathbb{R}^3).
\end{aligned}
$$

Hint. Take advantage of the fact that the solution to the problem is spherically symmetric, then $u(x_1, x_2, x_3, t) = u(0, 0, -|x|, t)$.

60) Solve the Cauchy problem for the wave equation

$$
\begin{aligned}
u_{tt} &= \Delta u + \sin t \sin(3x_1 + 4x_2) \quad (x \in \mathbb{R}^3,\ t > 0),\\
u|_{t=0} &= 0, \quad u_t|_{t=0} = x_2^2 + 4x_3^2 \quad (x \in \mathbb{R}^3).
\end{aligned}
$$

Hint. Use the superposition principle (linearity of the problem) and the fact that the function $\sin(3x_1 + 4x_2)$ is an eigenfunction of the Laplace operator.

61) Solve the Cauchy problem for the wave equation

$$
\begin{aligned}
u_{tt} &= \Delta u \quad (x \in \mathbb{R}^3,\ t > 0),\\
u|_{t=0} &= x_1^2 + 2x_3^2, \quad u_t|_{t=0} = \sin(x_1 - 3x_2) + \left[1 + (x_1 - x_2 + 3x_3)^2\right]^{-1} \ (x \in \mathbb{R}^3).
\end{aligned}
$$

Hint. Apply the superposition principle and reduce the problem to a lower-dimensional case.

62) Solve the Cauchy problem for the wave equation ($a > 0$ is a constant)

$$
\begin{aligned}
u_{tt} &= a^2 \Delta u \quad (x \in \mathbb{R}^3,\ t > 0),\\
u|_{t=0} &= \varphi(|x|), \quad u_t|_{t=0} = 0.
\end{aligned}
$$

Hint. Consider the function $v(r, t) = ru(r, t)$, $r = |x|$.

63) Solve the Cauchy problem for the heat equation

$$u_t = \Delta u \quad (x \in \mathbb{R}^n,\ t > 0), \quad u|_{t=0} = e^{-a|x|^2} \quad (x \in \mathbb{R}^n),$$

where $a > 0$ is a constant.

64) Solve the Cauchy problem for the heat equation

$$u_t = \Delta u + e^{x_1+x_2} \cos t \quad (x \in \mathbb{R}^2,\ t > 0),$$
$$u|_{t=0} = (3x_2 - 1)e^{-2x_2^2}.$$

Hint. Use the superposition principle and the fact that $e^{x_1+x_2}$ is an eigenfunction of the Laplace operator.

65) Solve the following problems for the heat equation on the semiaxis ($h > 0$ is a constant):

(a) $u_t = u_{xx} \quad (x > 0, t > 0), \quad u|_{t=0} = \varphi(x) \quad (x > 0), \quad u|_{x=0} = 0 \quad (t > 0);$

(b) $u_t = u_{xx} \quad (x > 0,\ t > 0), \quad u|_{t=0} = \varphi(x) \quad (x > 0), \quad u_x|_{x=0} = 0 \quad (t > 0);$

(c) $u_t = u_{xx} \quad (x > 0,\ t > 0), \quad u|_{t=0} = \varphi(x) \quad (x > 0),$
$(u_x + hu)|_{x=0} = 0 \quad (t > 0).$

Hint. Use a suitable extension of $\varphi(x)$ to $\mathbb{R}$.

66) Using semigroup theory, find an explicit form of a semigroup $T(t)$ generated by the operator $A : \mathcal{D}(A) \subset L_2(\mathbb{R}^n) \to L_2(\mathbb{R}^n)$,

$$Aw = \Delta w - a_0 w, \quad w \in \mathcal{D}(A) = H^2(\mathbb{R}^n), \quad a_0 \geqslant 0.$$

Hint. Obtain an explicit form of the resolvent of the operator A in Fourier images. Take advantage of the fact that the resolvent of the generator coincides with the Laplace transform of the operator-function $T(t)$.

67) Generalize the result given in Remark 19.1 as follows: if $f = 0$ and $\varphi \in H^{2m}(\mathbb{R}^n)$, then the classical solution u to problem (19.8), (19.9) belongs to the space

$$\bigcap_{k=0}^{m} C^k([0, +\infty), H^{2(m-k)}(\mathbb{R}^n)).$$

68) Find all values of a parameter $c \in \mathbb{C}$ for which the unbounded operator $A : \mathcal{D}(A) \subset L_2(0,1) \to L_2(0,1)$,

$$Au = -u', \quad u \in \mathcal{D}(A) = \{u \in H^1(0,1) : u(0) = cu(1)\},$$

is the generator of a contractive C_0-semigroup.

69) Find all values of a parameter $c \in \mathbb{C}$ for which the unbounded operator $A : \mathcal{D}(A) \subset L_2(Q) \to L_2(Q)$,

$$Au = c\Delta u, \quad u \in \mathcal{D}(A) = H^2(Q) \cap \mathring{H}^1(Q),$$

is the generator of a contractive C_0-semigroup. Here, Q is a bounded domain in $\mathbb{R}^n$ with smooth boundary.

70) Find all values of a parameter $c \in \mathbb{C}$ for which the unbounded operator $A : \mathcal{D}(A) \subset L_2(Q) \to L_2(Q)$,

$$Au = \Delta u + cu, \quad u \in \mathcal{D}(A) = H^2(Q) \cap \mathring{H}^1(Q),$$

is the generator of a contractive C_0-semigroup. Here, Q is a bounded domain in $\mathbb{R}^n$ with smooth boundary.

71) In the case of the third boundary-value problem in a bounded domain $Q \subset \mathbb{R}^n$,

$$\begin{aligned} \operatorname{div}(k(x)\nabla u) - a(x)u &= f \quad (x \in Q), \\ (\partial u/\partial \nu + \sigma(x)u)|_{\partial Q} &= 0, \end{aligned}$$

consider an unbounded operator $A : \mathcal{D}(A) \subset L_2(Q) \to L_2(Q)$ defined by the following formulas:

$$\begin{aligned} Au &= \operatorname{div}(k(x)\nabla u) - a(x)u, \\ u \in \mathcal{D}(A) &= \{u \in H^2(Q) : (\nu\nabla u + \sigma(x)u)|_{\partial Q} = 0\}. \end{aligned}$$

It is assumed that $\partial Q \in C^2$, $k \in C^1(\overline{Q})$, $k(x) > 0$ in $\overline{Q}$, $a \in C(\overline{Q})$, $a \geqslant 0$, and $\sigma \in C(\partial Q)$, $\sigma \geqslant 0$. Prove that the operator A is the generator of a contractive C_0-semigroup.

Hint. Apply the Hille–Yosida theorem, prove that the solutions of the operator equation $\lambda u - Au = f, \lambda > 0$ are weak solutions of the third boundary-value problem. In order to estimate the resolvent of the operator A, use the corresponding integral identity.

List of symbols

https://doi.org/10.1515/9783112229637-007

Index

https://doi.org/10.1515/9783112229637-008

Bibliography

[1] V. S. Vladimirov et al.: *A Collection of Problems on the Equations of Mathematical Physics*, Fizmatlit, Moscow, 2003 (fourth edition, in Russian); English translation: Mir Publishers, Moscow, Springer-Verlag, Berlin–Heidelberg, 1986.

[2] N. Dunford, J. T. Schwartz: *Linear Operators. Part 2. Spectral Theory*, Wiley-Interscience, New York, 1963.

[3] A. M. Ilyin: *Equations of Mathematical Physics*, Fizmatlit, Moscow, 2009 (in Russian).

[4] K. Yosida: *Functional Analysis*, in Classics in Mathematics Series, Springer-Verlag, Berlin–Heidelberg, 1995 (6th edition).

[5] A. N. Kolmogorov, S. V. Fomin: *Elements of the Theory of Functions and Functional Analysis*, Nauka Publishers, Moscow, 1976 (fourth edition, in Russian); English translation of the first 1957 edition in Dover Books on Mathematics, Dover Publications, Mineola NY, 1999.

[6] M. A. Krasnosel'skii: *Topological Methods in the Theory of Nonlinear Integral Equations*, Gostekhisdat, Moscow, 1956 (in Russian); English translation: International Series of Monographs on Pure and Applied Mathematics, Vol. 45, Pergamon Press, Oxford–London–New York–Paris, 1964.

[7] R. Courant: *Partial Differential Equations*, in Methods of Mathematical Physics by R. Courant and D. Hilbert, Vol. 2, Interscience, New York, 1963.

[8] O. A. Ladyzhenskaya: *The Boundary Value Problems of Mathematical Physics*, Nauka, Moscow, 1973 (in Russian); English translation: Applied Mathematical Sciences, Vol. 49, Springer-Verlag, New York–Berlin–Heidelberg–Tokyo, 1985.

[9] J. L. Lions, E.Magenes: *Non-Homogeneous Boundary Value Problems and Applications. Vol.* 1, Springer-Verlag, Berlin–Heidelberg–New York, 1972.

[10] S. Mizohata: *The Theory of Partial Differential Equations*, Cambridge University Press, London, 1973.

[11] V. P. Mikhailov: *Partial Differential Equations*, Nauka, 1976 (in Russian); English translation: Mir Publishers, Moscow, 1978.

[12] S. G. Mikhlin: *Linear Equations of Mathematical Physics*, Nauka, Moscow, 1964 (in Russian); English translation: Holt, Rinehart and Winston, New York, 1967.

[13] O. A. Oleinik: *Lectures on Partial Differential Equations*, Moscow State University Press, Moscow, 1976 (in Russian).

[14] W. Rudin: *Principles of Mathematical Analysis*, McGraw-Hill, New York–Toronto, 1976 (third edition).

[15] W. Rudin: *Functional Analysis*, McGraw-Hill, New York–Toronto, 1973.

[16] S. L. Sobolev: *Application of Functional Analysis in Mathematical Physics*, Leningrad St. Univ. Press, Leningrad, 1950 (in Russian); English translation: AMS, Providence, 1953.

[17] S. L. Sobolev: *Partial Differential Equations of Mathematical Physics*, Gostekhisdat, Moscow, 1947 (in Russian); English translation: Pergamon Press, Oxford–London–New York–Paris, 1964.

[18] V. A. Solonnikov, N. N. Uraltseva: *Sobolev Spaces*, in: Selected Topics of Higher Algebra and Analysis, Leningrad, Leningr. Gos. Univ. 1981 (in Russian).

[19] E. Stein: *Singular Integrals and Differentiability Properties of Functions*, Princeton University Press, 1970.

[20] A. N. Tikhonov, A. A. Samarskii: *Equations of Mathematical Physics*, Moscow University Press, Moscow, 1999 (6-th edition, in Russian); English translation: Dover Books on Physics, Dover Publications, Mineola NY, 2011.

[21] A. S. Shamaev et al.: *A Collection of Problems on Partial Differential Equations*, Binom Press, Moscow, 2009 (second edition, in Russian).

[22] M. A. Shubin: *Lectures on Equations of Mathematical Physics*, Moscow Center for Continuous Mathematical Education, Moscow 2001 (in Russian).

[23] R. A. Adams: *Sobolev Spaces*, Academic Press, New York, 1975.

[24] L. C. Evans: *Partial Differential Equations*, in Graduate Studies in Mathematics, Vol. 19, AMS Press, Providence RI, 2010 (second edition).

[25] A. Pazy: *Semigroups of Linear Operators and Applications to Partial Differential Equations*, Springer-Verlag, New York, 1983.

https://doi.org/10.1515/9783112229637-009

[26] L. E. Rossovskii, A. L. Skubachevskii: *Partial Differential Equations. Pt. 1: Function Spaces. Elliptic Problems*, RUDN University Press, Moscow, 2015.

[27] L. E. Rossovskii, A. L. Skubachevskii: *Partial Differential Equations. Pt. 2: Evolutionary Equations*, RUDN University Press, Moscow, 2015.

www.ingramcontent.com/pod-product-compliance
Lightning Source LLC
Chambersburg PA
CBHW081522220326
41598CB00036B/6296

* 9 7 8 3 1 1 2 2 2 9 6 2 0 *